Part One —Number Lines   Do you know who we are?

Our Mathematics Book

to order this book, the Link is: http://www.amazon.com/dp/1546548174

Only the Questions, not the problems, have
answers in this book with full explanations.

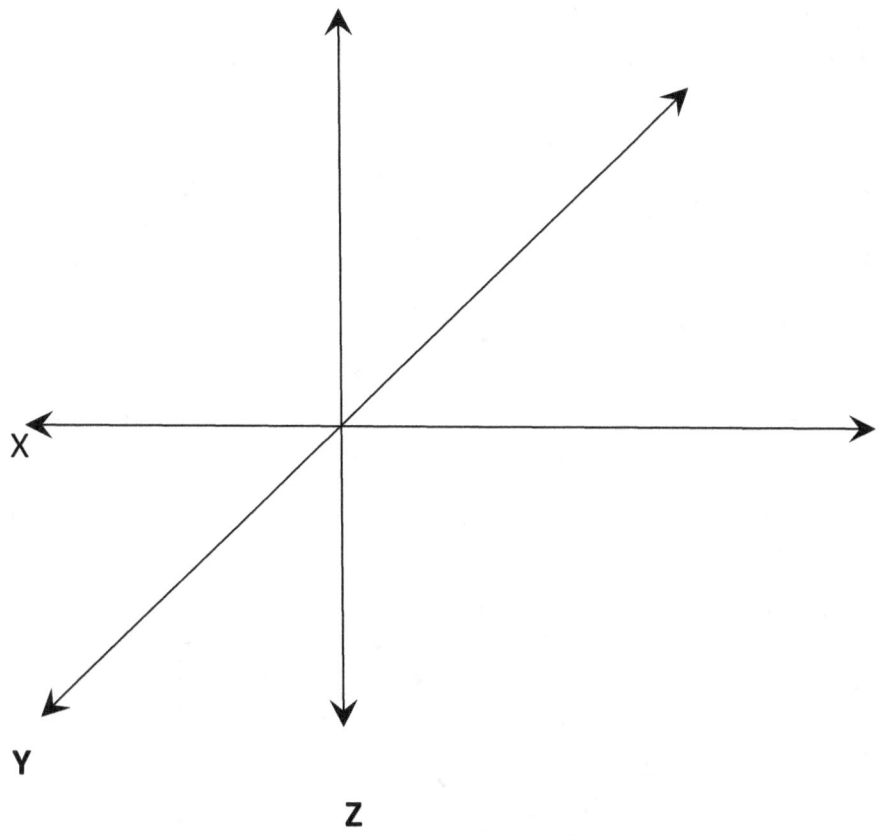

These are the famous X, Y, Z lines. Line X represents the left-right direction, line Y represents the forward-backward direction, and line Z represents the up-down direction, in other words it is a direction that starts below the ground and ranges beyond the sky. These lines have an infinite length. They all intersect the origin in a manner that causes the creation of a 90° angle between each of the three lines.

The slope of a function is the "Rise" over the "Run." This is what occurs with a XY graph. The graph either rises or falls or the line moves towards the top or the bottom of the page as X does a "run" or progresses towards the right hand side of the graph. The number next to the letter Y tells us how much Y changes for every unit, where unit means the number one in most cases, movement of X. Where the function intersects with the Y axis, also known as the Y intercept, is where the value of X has a zero value.

The slope of a function is found in a formula. The formula has the format of mx + b where m stands for the slope of the function and b represents the Y intercept.

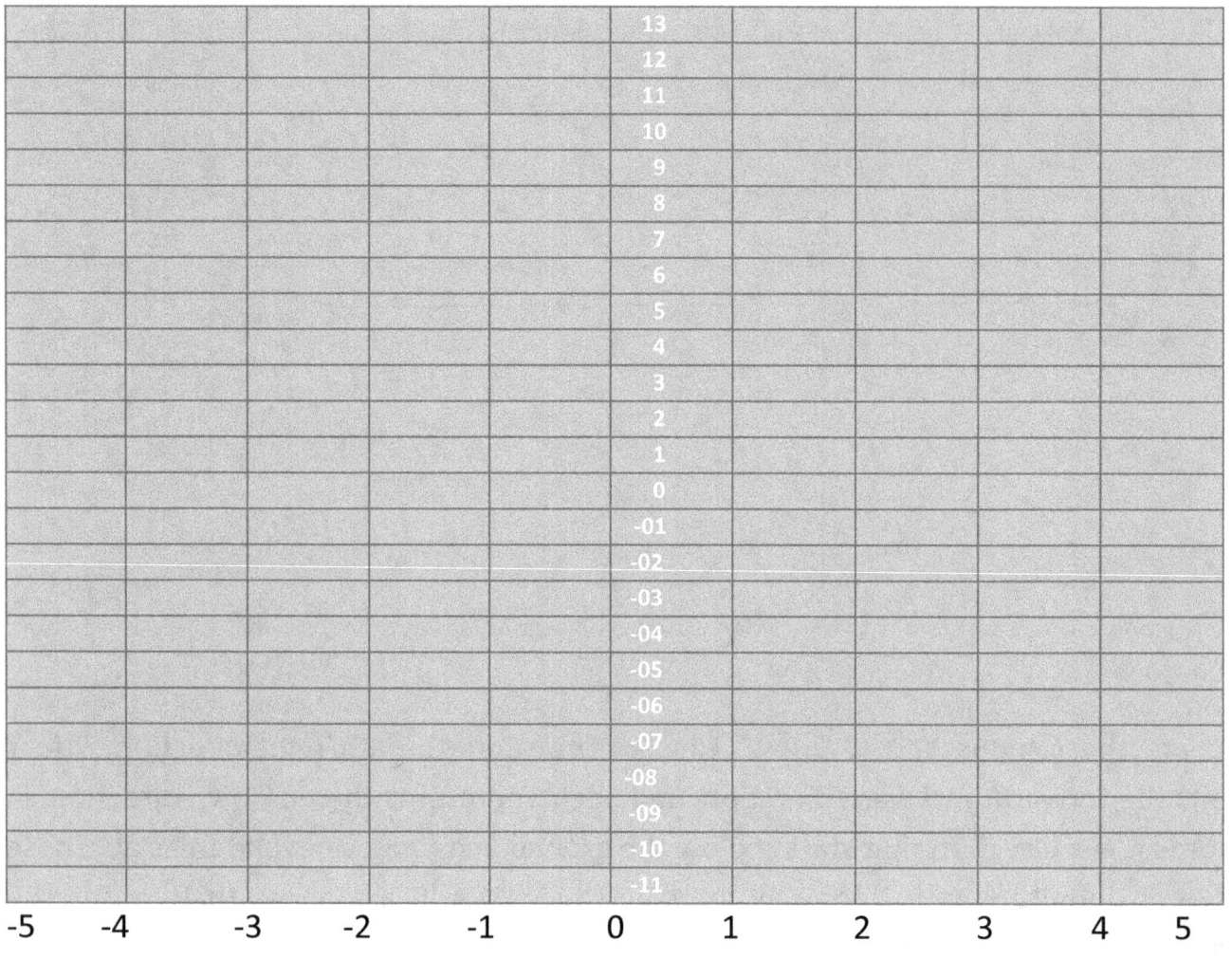

Ragtime Music

We will start with the X and Y axes. The horizontal zero line, or the X axis, is indicated by the number zero in the far right column. The vertical zero line, or the Y axis, is the line that has a zero under it and this zero is in between two number ones. Although the above graph has no arrows, arrows represent the direction that numbers progress in an increasing manner of a line that is continuous.

					13					
					12					
					11					
					10					
					9					
					8					
					7					
					6					
					5					
					4					
					3					
					2					
					1					
					0					
					-01					
					-02					
					-03					
					-04					
					-05					
					-06					
					-07					
					-08					
					-09					
					-10					
					-11					

-5   -4   -3   -2   -1   0   1   2   3   4   5

Part One Problems

Graph the two functions.

1A.) $X^2 + 0.5 = Y$

1B.) $-X + 2X = Y$

1C.) $Y(-4) = 2X$

1D.) (More Challenging)

X times Y equals one half. Graph this equation.

1E.) $X + 2 = 6$

1F.) $X - 6.5 = Y$

Part Two

All operations start at the origin, <u>also known as (0, 0) or zero.</u> For example,

should we move from the origin five integers, whole numbers such as 1, 2, 3, 4, 5, etc., to the left we will arrive at -5. What happens if we move fifteen spaces to the left from there? Assume that this graph had many more spaces or lines indicated on it than it does, now. If you think that I am moving slow, good. The way toward understanding is to use "simplicity." Here, it is better for us to ask the complete question of if we move five integers in a negative direction and then moved fifteen integers in a negative direction, where will we be on the X axis? This is good wording because we start at the origin and did not take a chance at confusing ourselves with starting at negative five and wondering whether we drew adding or subtracting a movement of fifteen integers. So, again, first we moved in a negative direction from the origin for a movement of five integers followed by a movement in the negative direction of fifteen integers. The result is negative twenty. Had the question been what is the result of -5 + 15 then we would had, again, started from the origin to move five integers to the left and then we would of moved fifteen integers to the right. Here, the result would be ten integers right of the origin.

Again, all operations begin at the origin. There are short cuts one can use. They work well but by starting at the origin the correct answer is usually reached and noticed.

Movements are represented by signs such as + for a positive movement or – for a negative movement. Here, positive is "understood" to be a right side movement and negative is understood to mean "the other way, or direction." So, if one tells you to move three, or a positive 3, you are to move from zero to positive three but if he tells you to move a negative three spaces you are to move three spaces to negative three. If he tells you to add a "negative negative" three you are to start at the origin and move "the other way" "the other way." The phase "the other way" was printed twice because the word negative appeared twice. First you prepare to interpret "the other way" to mean move towards the left but then the second negative sign or phase "the other way" tells you don't move in that "other way" but move in an opposite "other way." Here, you will move three integer spaces to the right because two negatives, grammatically, equal a positive.

100	82	-202	- 8.0
16	3	791	20.0
- 60	- 66	-811	19.5
56	19	-222	31.5

Here is a fictional story to help you/us with mathematics. You had a birthday party and invited my Aunt Sally. She got drunk and started throwing food and vomit at your guest. To calm her down one of the guests gave her a job. Ok, my aunt, she got a job in a "bedroom community" as a labor who cuts grass off of people's lawns and off of an ice cream factory's lawns.

At work, she did the wrong lawns in one block and the lawn mower blew a fuse that jumped in the air and cracked an ice cream truck's rear view mirror. She, Aunt Sally, blamed the truck driver for driving illegal ice cream. To escape her "non-sense" the driver gave her some sandwiches and a milkshake. She got on the bus with the food and spilled a cold soda and the ice cream sandwich on the bus drivers' shoes because she didn't like the order of his operations. He, the bus driver, called 9-1-1 to arrest her. Now, she needs you, again, to get her out of trouble by remembering the order of operations.

They are "Please Excuse My Dear Aunt Sally."

This represents first doing work that is between parenthesis, second doing work of exponential operations, third doing division and multiplication operations, and last doing addition and subtraction operations.

Example: $(2)(78/3 + 18 \cdot 5/9 - 27)^2 + (10.2 \cdot 2)(8 - 4)/2$
Here, a dot between numbers, such as $18 \cdot 5$, means multiply the two numbers and where two sets of parenthesis are next to each other, such as $(10.2 \cdot 2)(8 - 4)$, the value of what is in the first parentheses is to be multiplied by the value of what is in the second parentheses.

Part Two Problems

2A.) Add 34.009 plus 45.000367. What is the result of this combination?

2B.) Combine these two numbers, 245.00449 and -544,383.36605.

2C.) Combine these two numbers, -5.0023089 and -54,383.06500.

Combine through Addition/Subtraction *(warning: Be careful)*

2D.)	467.0998	2E.)	4,009.889	2F.)	-230.0998	2G.)	22.0009190
	458,0099		-62,308,296		-003.669		- 0.0000059059
	-0.07709		50,0503.00		-00900.086.		

2H.)	160.0908	2I.)	-230.0998	2J.)	-228,648,200.58
	+ 0.07709		-258,001 .		50,120,030.35

2K.) -23.0098 - 455 = ?

Part Three

What are Fractions?

On each number line there are an infinite number of points and points between points. Wow!

A   B

So, the number of points between A and B is infinite. Let's call A zero and let us call B one. Let's look at some, not all, of the points between or at A and B.

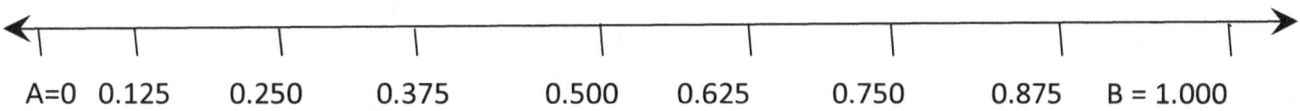

A=0 0.125   0.250   0.375   0.500   0.625   0.750   0.875   B = 1.000

In the second number line we see eight sections so the difference between A and B, which is one, has to be divided by eight. 1/8 = 0.125. Because the result is 0.125 it should be added eight times to A. Had we decided to have only five sections then each section would have had only a value of 0.2 because

$$1/5 = 0.200.$$

A = 000   0.200   0.400   0.600   0.800   B = 1.000

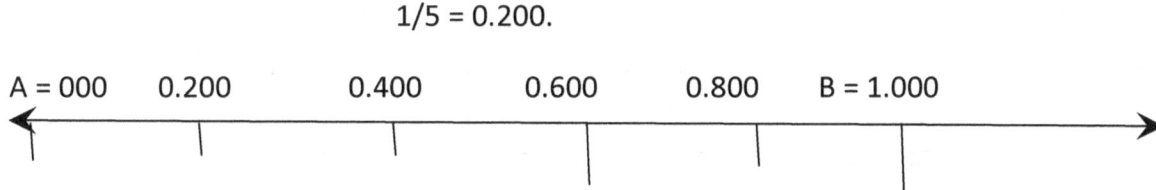

George Washington Carver

Fractions Simplified

1.)

A = 0.00   0.500   B = 1.000

2.)

A = 0.000   0.333   0.666   B = 1.000

3.)

A = 0.000   0.250   0.500   0.750   B = 1.000

4.)

A = 0.00   0.200   0.400   0.600   0.800   1.000

5.)

A = 0.00   0.166   0.333   0.500   0.666   0.833   B = 1.000

6.)

A = 0.0000   0.1428   0.2856   0.4284   0.5712   0.7140   0.8568   B = 1.0000

Public Enemy

To find the size of each interval, divide the difference between A and B by the number of spaces indicated on each of the graphs, 1 to 6. In number line 5 we see a few numbers with a line over it. The line represents an infinite repetition of the numbers under that line.

A Family Reunion

By the O'Jays

The spaces of a decimal number right of the decimal point have names. The first space is the tenths followed by hundredth, thousandths, ten thousandths, hundred thousandths, millionths, etc., which represent the fact that we divided the number one by ten, one hundred, one thousand, ten thousand, one hundred thousand, one million, etc. In number line six the number that is one space left of point 1.0000 is 0.8568 or eight thousand five hundred and sixty-eight ten thousands parts of one. In other words if we divided the difference between A and B into ten thousand equal parts the number 0.8568 would represent 8,568 of those ten thousand, 10,000, parts. In number line one the number that is one space left of point 1.00 is 0.5 or five tenths of one. In other words, if we divided the difference between A and B into ten equal size parts the number 0.5 would represent five of those ten parts.

Do not let the zeros, which follow an integer, "scare" you. They, or it if only one exists, represent a sound not much of any necessary meaning. Here, 0.5 represents 5 tenths parts, 0.50 represents 50 hundred parts, and 0.500 represents 500 one thousands parts. In all of these examples the numbers represent one half of the possible parts whether it was 5 of 10, 50 of 100, or 500 of 1,000.

Notes:

Part Three Problems

(Do all multiplication, division, addition, and subtraction without use of a calculator.)

3A.) How many 0.250 parts are there in 12?

3B.) How many 0.16 parts are in 725?

3C.) How many one halves are in two hundred?

3D.) Name this number 0.00030.

3E.) Name this number -34.003.

3F.) Change 0.00908 percent into a decimal number.

3G.) Change 1000 in to a percentage number.

3H.) What percentage number represents one whole?

3I.) Write 200 percent as a decimal number.

3J.) Is 4/8 greater than, less than, or equal to 101/102?

3K.) How many 0.250 parts are there in 12?

3L.) How many 0.48048 parts are in 1550?

3M.) How many one halves are in two thousand?

3N.) Name this number 0.44030.

3O.) Name this number -288,124.00345.

3P.) Change 0.0098 percent into a decimal number.

3Q.) Write 10.00 as a decimal number.

3R.) What percentage number represents one-half?

3S.) Write 200 percent as a percentage number.

3T.) Is 16/10 greater than, less than, or equal to 101/102?

15

Notes

More **Notes**

So, 6/8 is what? It is the number one divided into eight parts and then multiplied by six. Another way of saying what is 6/8 is to call it six of eight parts.

24 45	2	2
3	33	333
554	35	5 w3
654	53	5
754 T	535	4 4q
7 5	54546	46y6

Here, six of the eight above rows have numbers placed in them.

Six over eight equals 0.75., which is seventy-five equal size parts of a whole thing. When the decimal point is moved two spaces to the right the number is a percentage, 75%. Don't let us fool you by giving complicated questions such as write 00.00709 as a percent. 'The answer is found simply by moving the decimal point two spaces towards the right. 0.709% is the correct answer. Now, when you are to change a PERCENTAGE NUMBER, such as 82.097, IN TO A DECIMAL, YOU SHOULD MOVE THE POINT TO THE LEFT. Here, we will reach 0.82097. To change this number into a fraction, we find that it has five decimal spaces so the denominator has five zeroes after a one and the answer is 82097/100,000.

In the chart above six of the eight parts have numbers in them. The fraction 6/8 has a numerator and a denominator. They are separated by a division line. So, the numerator is to be divided by eight. Six over eight equals 0.75, which is 75 of one hundred equal size parts of a whole thing.

Summertime by Sam Cooke

Part Four

How do You Add (a simple question and a long answer)

I will skip the jargon and jump to a conclusion. Let's call adding and subtraction combining. Note, multiplication and subtraction are quick ways of adding or subtracting. Now the question is how do you combine? First, go to the origin and second move as you were shown a few pages above.

But, how do you add, I mean subtract, I mean combine fractions? That's a good question. If, the denominators of the numbers are the same then one only needs to add the numerators and keep the same denominator.

Examples

5/10 + 65/10 - 83/10 = -13/10

43/8 − 600/8 = -557/8

Of course -13 divided by ten is -1.3 and -557 divided by 8 is -69.625.

How do you combine two numbers that do not have common denominators?

Example;   (25/180 + 16/90) =?

Here, you, we, must get a common denominator. This is not necessary when you are combining numbers by multiplying them. When combining with the use of addition or subtraction, which are words I do not like to use but many of you do like them and may prefer to use them as they confuse the "exponents" out of your horizons, to get a common denominator you must find a number that both fractions' denominator can go into evenly. Here, evenly means integer, a number that has only zeroes on the right side of the decimal point and, here, evenly does not mean the opposite of oddly. Both 180 and 90 can be divided into 180. After finding 180 to be the new denominator we must multiply the number of times that the old denominator number went into this common denominator and multiply the fraction's numerator by the amount of times that the old denominator went into the new denominator. Here, 180 went one time into 180. So we are to multiply 25 times one to make a numerator part which is 25. 90 went into 180 two times so we then must multiply 16 times two to make a numerator part that is 32. The result is [(25 + 32)/ 180)] or (57/180).

This can be "reduced," where reduced does not mean a lower value but the fraction can receive smaller integers. Here, we can divide both the numerator and the denominator by a digit. Here, that digit can be three. Fifty-seven divided by three is nineteen and one hundred and eighty divided by three is sixty. So, the reduced, or new, fraction is nineteen over eighty, 19/60.

Part Four Problems

4A.) Multiply 0.008 by a negative 0.25.

4B.) Divide 0.16 by 5.

4C.) How many one halves are in 16?

4D.) Write negative fourteen and two-half as a decimal number.

4E.) Name this number 3.003.

4F.) Change 10.009 percent into a decimal number.

4G.) Add -1000 to 200,202.

4H.) What percentage number represents three wholes?

4G.) Write 4.5 times 4.6 as a percentage number.

4H.) Multiply 4/8 times 101/102.

4I.) Divide 0.250 by 200 and then multiply the result by 112. What is the result?

4J.) How many 10 size parts are in 1550?

4K.) How many one quarters are in two thousand-five hundred?

4L.) Name this number 04.030.008.

4M.) Name this number 448,004.3053.

4N.) Change 0.98 percent into a decimal number.

4O.) Write 0.4500 in to as a percent.

4P.) What percentage number represents 20.688?

4Q.) Write 0.02100 percent as a percentage number.

4R.) Is 6/10 greater than, less than, or equal to 1/102?

Part Five

Time for a Tangent & Fractions

Multiplication, division included, is a more complicated method of combining numbers than is addition, subtraction included.

The number three, 3, is the same as 3/1 which is the same as 9/3 which is the same as 27/9 which is the same as 81/27 which is the same as 243/81, etc. "Reducing a fraction" does not mean to reduce its value but to only make its integers, one of the numerator and one of the denominator, smaller. Let us look at (57/3)(3/180). Here 57/180 equals 19/60 after we multiply 57/180 by 3/3 or the number one. Multiplying anything, 57/180 included, by one does not change a thing. So, (57/180)(1/1) = (57/180), (57/180)(2/2) = (57/180)(3/3), and (57/180)(4/4) = (57/180)(X/X). Note, I do not use x or X for multiplication although many, including college, texts use the symbol x or X for multiplication. I prefer to use parenthesis or a dot, not a period, to indicate multiplication.

Going back to (57/3)(3/180), we can see that (57/3) = 19/1 and (3/180)(1/60) and (19/1)(1/60) is (19/60). Again, 57/180 = 19/60 and this was found by dividing both the numerator and the denominator by three.

(This is the end of the tangent.)

To multiply numbers that do or do not have common denominators just multiply the numerators and multiply the denominators.

Example: (208/1.77)(100/2.00) = 20,800/ 3.54 = 20,800 divided by 3.54. This is also true for division, except for, in division we are to invert the divisor and multiply it by the numerator.

Example: (10/2)/(16/4) = (10/2)(4/16) = 40/32

40/32 can be reduced to two numbers, one a numerator and one a denominator, that are 5/4. Here, both 40 and 32 were divided by eight. The result is a fraction that is greater than one. So, 5/4 = (4+1)/4 or 4/4 + 1/4. Here, 4/4 equals one. So the answer to (10/2)/(16/4) is 1 and one-quarter. We also could of noticed that with (10/2)(4/16) the number two can go into the number four two times and it goes into itself one time. The two fractions, thus, become (10/1)(2/16) which equals 20/16. After dividing both the numerator and the denominator by four the result is the fraction 5/4, which was reduced above to being 1 and 1/4.

Mama

Part Five Problems

(Do all multiplication, division, addition, and subtraction without use of a calculator.)

5A.) How many 0.50 parts are there in 212?

5B.) How many 0.160 parts are in 25?

5C.) Divide -22 by -02000

5D.) Name this number11 0.0220.

5E.) Name this number 04.003.

5F.) Multiply 0.0330908 by 602.009.

5G.) Divide 1000 by 725.0725.

5H.) What percentage number represents sixteen?

5J.) Divide 200 by 16/18.

5K.) Multiply 4/8 by 01.102.

5L.) Divide 0.250 to the fourth power by 16 cubed.

5M.) Multiply -0.48 by -160.

5N.) Divide thirty-three by two thousand?

5O.) Name this number 0.4880.

5Q.) Name this number -20,345.010122.

5R.) Multiply 0.0098 percent by 300.

5S.) Divide 1,210.00 by 10,022.40.

5T.) What percentage number represents one-half squared times one-halve cubed?

5U.) Write 0 percent as a percentage number.

5V.) Is 1/10 greater than, less than, or equal to 1/02?

Part Six

Thinking of Division Multiplication is the inverse of division and division is the inverse of multiplication. Example: $(6/1 \cdot 3/1) = (18/1)$ or six times three is 18. Also, $(6/1 \cdot 1/3) = (6/3)$, which is six divided by 3 and equals two.

This is why when we divide two fractions such as 3/9 divided by 33/18 we invert the divisor and then multiply. An example of this is what happens when we divide 3/9 by 33/18.

$(3/9)(18/33)$ is the same as $(3/33)$ - which is a fraction that can be reduced to being 1/11 - times $(2/1)$. This becomes, after multiplying, $(2/11)$. Here, three was divided by THREE to give a result of one and thirty-three was divided by that same THREE to give a result of eleven. Nine was divided by NINE and 18 was divided by that same NINE to give a result of two. The resulting "reduced" fractions were multiplied to give us a result of 2/11 or two divided by eleven.

Below is a picture of the Edmund Pettus Bridge where Dr. Martin Luther King, Jr. was attacked on Bloody Sunday by legal armed forces and below that is a picture of the White House.

Long Division Procedures

What is twenty-four and two thousand four ten thousandths divided by eight tenths?

$$60.234\overline{)75.894840}$$

1.) Move the decimal point of the divisor to the far right.

2.) Move the decimal point of the dividend the same number of spaces to the right.

3.) Place a decimal point just above the top of the radical sign.

4.)

```
              1.
        _____
60234. | 75894.84
        -60234
        _____
         15660
```

60,234 goes one time into

75,894 one time and one times 60,234 taken from 75,894 should be taken 15660.

5.) Bring down the next digit, an eight, 8.

                1.2   60234 goes two times into 156,608 and two times

60234. $\overline{75894.84}$ 60234, or 120,468, should be taken from 156,608.

          -60234

          156608

6.) Bring down the next digit, a four, 4.

                1.2   60234 goes two times into 156,608 and two times

60234. $\overline{75894.84}$ 60234, or 120,468, should be taken from 156,608.

          -60234

          156608

        -120468

7.)   361404   Bring down the next digit, a four, 4.

                        60234 goes into 361404 six times.

              1.26

60234. $\overline{75894.84}$

          -60234

          156608

       - 120468

          361404   Six times 60234 from 60234 ends the computation.

         -361404

             0

The Edmond Pettus Bridge of Selma, Alabama and the Capitol Building of Washington District of Columbia

Part Six Problems

(Do all multiplication, division, addition, and subtraction without use of a calculator.)

Sarah Boone and the Ironing Board

6A.) How many 0.750 parts are there in -21?

6B.) How many 0.160 parts are in 25?

6C.) Divide -22 by -02000

6D.) Name this number11 0.0220

6E.) Name this number 04.003

6F.) Multiply 0.0330908 by 602.009.

6G.) Divide 1000 by 725.0725

6H.) What percentage number represents sixteen?

6I.) Divide 200 by 16/18.

6J.) Multiply 4/8 by 01.102.

6K.) Divide 0.250 to the fourth power by 16 cubed.

6L.) Multiply 0.48 by -1550^2.

6M.) Divide thirty-three by two thousand?

6N.) Name this number 0.4880

6O.) Name this number -20,345.010122

6P.) Multiply 0.0098 percent by 300.

6Q.) Divide 1,210.00 by 10,022.40.

6R.) What percentage number represents one-half squared times one-halve cubed?

6S.) Write 0 percent as a percentage number.

6T.) Is 1/10 greater than, less than, or equal to 1/02?

Part Seven

$2^4 = 16$ or $(2)(2)(2)(2)$. This is also $(2^1)(2^1)(2^1)(2^1)$ but usually the exponent of one is so understood that it is rarely written. When multiplying we are to add exponents and when dividing we are to subtract exponents.

$4^6/4^2 = 4^4$. This is the same as $(4)(4)(4)(4)(4)(4)/(4)(4)$

The two fours of the denominator cancel out two of the fours that are in the numerator. This leaves a remainder of four number fours in the numerator which is the same as 4^4.

$2^4 = 16$ or $2^{4/1} = 16$. The denominator of the exponent tells us the route of the number. Here, the route is one. What about $5^{4/2}$, what does it tell us? Here, 5 to the 4th power is 625 and five squared, or to the second power, is 25. By dividing 625 by 25 we see a result of 25. In other words by canceling we could of changed the 4/2 fraction to being 2/1 and found that 5 to the second power is twenty-five. This is the same as $(5 \cdot 5 \cdot 5 \cdot 5)/(5 \cdot 5)$. The two fives in the denominator would cancel two of the four fives that are in the numerator and the result would be five squared or 25.

Let's look at $(4)^{6/2} = (4)(4)(4)(4)(4)(4)/(4)(4) = (16)(16) = 256$.

This pattern can work when the exponent has a denominator of two. So, it is best to reduce the exponent's fraction to another form.

Example: $(64^{2/4}) = (64^{(1/4-2)}) = 64^{1/2} = 64^{0.5}$ and this equals eight.

It is time to subtract, or add, integer values of exponents when the exponents are on opposite sides of a fraction. Here, whatever you take off of the top you must take off of the bottom.

Examples: $16^4/16^3 = 16^3/16^2 = 16^2/16^1 = 16^1/16^0 = 16$

$16^3/16^4 = 16^2/16^3 = 16^1/16^2 = 16^0/16^1 = 1/16 = 0.0625$.

Placing something to the one-half power is the same as getting that something's square root. Placing something to the one-quarter power is the same as getting that something's quarter root. Anything to the third power is the same as anything cubed. Anything raised to the second power is the as getting that anything squared.

Is there a difference between $(16)^{1/2}$ and $(16^{1/2})$?

Notes

Notes

In my little Town   By Simon and Garfunkel

In my little town
I grew up believ--ing
God keeps his eye on us all
And he used to lean upon me
As I pledged allegiance to the wall
Lord I recall
My little town

Coming home after school
Flying my bike past the gates
Of the factories
My mom doing the laundry
Hanging our shirts
In the dirty breeze

And after it rains
There's a rainbow
And all of the colors are black
It's not that the colors aren't there
It's just imagin-ation they lack
Everything's the same
Back in my little town
Nothing but the dead and dying
Back in my little town
Nothing but the dead and dying
Back in my little town

In my little town
I never meant nothin'
I was just my fathers son
Saving my money
Dreaming of glory
Twitching like a finger
On the trigger of a gun
Leaving nothing but the dead and dying
Back in my little town
Repeat and fade:
Nothing but the dead and dying
Back in my little town

Now I know what you're thinking. You want to know what if, or something. What happens, here, if we have a number raised to an exponent that is raised to another exponent such as

$$[X^{(5 +b)\wedge 3}]?$$

Here the three that follows the ^ symbol is the exponent of $(5 + b)$. Here we must multiply $(5 + b)$ by its self three times. The first time is just $(5 + b)$. The second time is $(5 +b)(5 + b)$. The third time is $(5 + b)(5 +b)(5 + b)$. The FOIL method of multiplication will be used where FOIL stands for First, Outside, Inside, and Last. Here, we multiply the two first numbers of each parenthesis, sum that to the product of the two outside numbers of each parenthesis, add that to the product of the inside numbers of the two parentheses, and finally add that to the product of the two last terms of the two parentheses.   This is $(5)(5) + (5b) + (b)(5) + (b)(b) = 25 + 10b + b^2$. Next, the result is to be multiplied by $(5 + b)$. Here, $(25 + 10b + b^2)(5 + b)$ is equaled to $125 + 50b + 5b^2 + 25b + 10b^2 + b^3$. After unscrambling the terms we have $b^3 + 15b^2 + 75b + 125b^0$, or $b^3 + 15b^2 + 75b + 125$.   Note that the unknown, here b, is listed in a decreasing power order from as large as necessary down to the zero power. Since b^0 equals one and since b^1 is equaled to itself these two exponents do not have to be written so the answer is, again.

$$b^3 + 15b^2 + 75b + 125.$$

Part Seven Problems

*

(Except for questions 7Q and 7R,do all multiplication, division, addition, and subtraction without use of a calculator.)

7A.) Multiply 0.48 by -15.50^2.

7B.) Divide 0.750^4 by -21^4.

7C.)How many 0.16^2 parts are in 25^2?

7D.) Divide -22^4 by -020.

7E.) Name this number 10.022^{-2}

7F.) Divide 4.003^{16} by $1.04/8.080^2$ and place only four decimal places in the answer.

7G.) Multiply 0.0908^{-2} by 2.09^{-2}.

7H.) Divide 104 by 725.0725/100.

7I.) What percentage number represents ten thousand^2?

7J.) Divide 200 by 1/1.

7K.) Multiply 3/8 by 01.88.

7M.) $0.20^3 / 6^4$.

7N.) Multiply 0.8 by 15^3.

7O.) Divide thirty-eights by two hundred and forty-one.

7P.) Divide -0.48 by -16.

7Q.) Raise 10.03 to the 2.02 power.

7R.) Multiply 0.1098 percent by 300 to the -0.3 power.

7S.) $1.1001/(10,022.40/1.1001^3)$.

7T.) What percentage number represents one-halve cubed?

7U.) Write 0 percent as a percentage number.

7V.) Is 144,410 greater than, less than, or equal to 122,220/0.2?

33

Part Eight

A little more Work

$[(5 + X)/2(X-1) + 3X/(3-X)] = -4$.   The two denominators were multiplied together to create a new denominator. Each numerator was multiplied by the part of the new denominator that was not a part of its old denominator. The result is

$[(3-X)(5+X) + 3X2(X-1)]/2[X – 1](3-X)] =$ $15 -2X – X^2 + 6X^2 -6X/2[X – 1](3-X)] =$
$-8X + 5X^2 + 15/2[X – 1](3-X)]$ Multiplying
the right side of this equation by the new lefts side of the equation's denominator gives

$5X^2 – 8X + 15 = (-8X + 8)(3 – X)$

Using the foil method of multiplication give us a new format of the right hand side of the equation.

$5X^2 – 8X + 15 = -24X + 8X^2 + 24 - 8X$

Placing all terms on the left side of the equation creates what is next.

$-3X^2 + 24X – 9 = 0$ After dividing this equation by negative three it equals $X^2 - 8X +3 = 0$

To solve this one can "complete the square" or use "the quadratic formula" which is $\underline{-b +/- (b^2 -4ac)^{0.5}}$

$2a$.

Completing the Square

After doing the above, which includes subtracting three from the equation, to the equation take one-half of the X coefficient then square it and add the result to the other side of the equation. On the original side of the equation join it with X^2 but now house it in a set of parenthesis with the X of X^2 and place the exponent of X^2 on the right half of the parenthesis. Below is the result.

$(X -4)^2 = -3 + 16$

Next, find the square root of the equation, and this will create two answers, one a positive and one a negative answer on the right hand side of this equation.

$X – 4 = {}^{+/-} (13)^{0.5}$. So $X = 4 {}^{+/-} (13)^{0.5}$. This is the same as X = (7.605551275, 0.394448724).

The Quadratic formula To find when will (X + something) = 0 we can complete the square or use this formula.

$$\frac{8 \, ^{+/-}[(-8)^2 - 12]^{0.5}}{2} = 0.$$

Here, the letters of the quadratic formula represent the coefficients of the equation that is to be solved, X^2 − 8X + 3, where the letter a represents the one that is assumed to be next to the letter and power X^2, the letter b represents the negative eight that is next to the letter X, and the letter c represents the lone coefficient three, 3. Here, the first coefficient is given the letter a, the second coefficient is given the letter b, and the third coefficient is given the letter c.

After using the quadratic formula you will see that the results are (X + − 7.6055512285) and (X + - 0.394448725), or (X − 7.6055512285) and (X - 0.394448725). So, when X equals either 7. 6055512285 or 0.394448725 the equation will equal zero, 0.

Charles Drew (Who was he besides a good football player?)

The People Who Play Zydeco Music: To understand any genre of music, you must first understand the makers of that genre. Zydeco is the music of Southwest Louisiana's Black Creoles, a group of people of mixed African, Afro-Caribbean, Native American and European descent. This Creole society that beget zydeco is traditionally rural, French-speaking, and is somewhat intertwined with the Cajun culture.

Where Does Zydeco Come From?: Zydeco music is a relatively new genre of world music, having come about as a style of its own in only the mid-1900s. It is a derivative of "La-La" music (the shared music of the Cajuns and the Creoles), as well as blues, jure' (syncopated a cappella religious songs), and in more recent years, zydeco has taken many cues from R&B and even hip-hop, proving that it's a constantly evolving genre.

What Does "Zydeco" Mean? - Story #1: The word "zydeco" has two different stories to explain it. One is that it comes from the phrase "Les haricots sont pas sales", meaning "the snap beans aren't salty". This phrase is a colloquial expression meaning that times are hard, and when spoken in the regional French, it's pronounced "zy-dee-co sohn..." etc.

What Does "Zydeco" Mean? - Story #2: The second often-accepted meaning of the word "zydeco" is that it comes from the word "zari", which means dance. The word "zari" is found in several West African languages (in various similar forms).

Zydeco Instrumentation: Zydeco bands generally include an accordion, a modified washboard called a *frottoir*, electric guitar, bass and drums. Secondary zydeco instruments include fiddles, keyboards and horns.

What Does Zydeco Sound Like?: Zydeco music is often portrayed incorrectly as being polka-esque, but it actually sounds much more like the blues than like any European music. The band plays heavily on the backbeat, with modern bands relying on a double-kick to the bass drum to emphasize the syncopation. The accordion plays blues licks, and the guitars further emphasize this sound.

Zydeco Lyrics: Zydeco music is sung in both English and French, with English being the preferred language for most modern bands. Many zydeco songs are simply reworkings of R&B or blues songs, many are modern versions of very old Cajun songs, and many are originals. Song lyrics deal with everything from the mundane to intense socio-political issues, with food and love being two very common themes.

Clifton Chenier - The King of Zydeco: What Bill Monroe was to bluegrass, Clifton Chenier was to zydeco. He was the one who took zydeco from older "La-La" music to what we now recognize, and Clifton Chenier is hailed by nearly everyone as the progenitor of the modern genre.

Zydeco Dancing: Zydeco, like all accordion music, is for dancing. The steps performed to zydeco music look like swing dancing to those unfamiliar with it. Zydeco dancing is intensely passionate and sexy, and many are heralding it as "the new salsa."

Zydeco Music 101 By Megan Romer, About.com Guide

Some Factoring Facts

If the quadratic formula does not give you useable solutions try graphing the function. You will often see that the function does not reach a value of where Y has a value of zero.

Some common products used in factoring.

One.) A Perfect Square

$A^2 + 2ab + b^2$ becomes $(a + b)^2$

$A^2 - 2ab + b^2$ becomes $(a - b)^2$

Eg. $9X^2 + 30 + 25 =$

$(3X)^2 + 2(3X)(5) + 5^2$ and this becomes $(3X + 5)(3X + 5)$.

Two.) The difference of Perfect Squares

$X^2 - a^2$ becomes $(X - a)(X + a)$

Eg. $9X^2 - 4 =$

$(3X)^2 - (2)^2 = (3X - 2)(3X + 2)$

Three.) The sum of two cubes

$X^3 + a^3$ becomes $(X + a)(X^2 - aX + a^2)$

Eg. $64a^3 + b^3 = (4a + b)(16a^2 - 4ab + b^2)$

Four.)   The difference of two cubes

$X^3 - a^3$ becomes $(X - a)(X^2 + aX + a^2)$

One should notice some patterns. Observe these following patterns.

Difference of two squares $x^2 - a^2 = (x - a)(x + a)$

Also, $5^2 - 4^2 = 9$ or the sum of five plus four,

$8^2 - 7^2 = 15$ or the sum of eight plus seven,

$13^2 - 12^2 = 25$ or the sum of thirteen plus twelve.

Sum of two squares $x^2 + a^2 = (x + ai)(x - ai)$

This is similar to the "difference of two square," but here we will introduce a 'new' number. This number is called "the imaginary number, i." [Yes, we even have imaginary numbers, too.] It is equaled to the square root of negative one. So, i^2 equals -1. Looking at the sum of two squares, again, $x^2 + a^2 = (x + ai)(x - ai)$, with the foil method of multiplication we will see that the result will be $X^2 - [ai]^2$, which is the same as $X^2 - a^2(-1)$ and this will be $X^2 + a^2$. So, with the assistance of "I" we can factor the sum of two squares.

A sum of 2 cubes $x^3 + a^3 = (x + a)(x^2 - ax + a^2)$

A difference of 2 cubes $x^3 - a^3 = (x - a)(x^2 + ax + a^2)$

If the sign between the two terms that are in the left side of the equation is positive, the sign between the two terms of the first parenthesis set on the right hand side of the equal sign will be positive and the signs of the next set of terms that are in the second set of parentheses will, starting from positive to negative, alternate.

If the sign between the two terms that are in the left side of the equation is negative, the sign between the two terms of the first parenthesis set on the right hand side of the equal sign will be negative and the signs of the next set of terms that are in the second set of parentheses will be positive.

Here is an important pattern to notice. The exponent that is on the left side of the equal sign will be one integer greater than the largest exponent that is on the left hand side of the equal sign. Note that on the right hand of the equal sign there are no exponents greater than one except for being within the second set of parentheses. There, the greatest exponent is one integer less than the exponent value that is on the left hand side of the equal sign. Moving from the left to the right, each unknowns' exponent value, on the right hand side of the equal sign, will have a value that is one integer less than it previously had when that unknown was previously written. This will continue until a value of zero is the exponent's value in the last term of the set of terms that are within the last parenthesis that are on the right hand side of the equation. On the other hand, moving from the right to the left, on the right hand side of the equal sign, unknowns' exponent value will have a value that is one integer less than it had when that unknown was written as the term that is towards the left. This will continue until a value of zero is the exponent's value in the last term of the set of terms that are within the last parenthesis that are on the right hand side of the equation.

 This pattern continues for all exponents.

Going further we will notice a pattern of exponents and coefficients.

When the different two unknowns that have a similar exponent are expanded the result will include the two different unknowns in one parenthesis set without any exponents other than the number one. These last results are multiplied by the last set of terms in a set of parentheses where one of them has an exponential value of being one less than the exponential value it had before they were expanded while the other is multiplied on to it as it has an exponential value of zero. Each following term of this final set of parenthesis will follow a pattern where the unknown that appeared in this parenthesis while having an exponential value of one less than what it had before the expansion will receive an exponent that is one less than the exponential value that it had previously until the last term is reached where this unknowns exponential value will be zero. On the other hand, the unknown that had, at the beginning of this second parenthesis, an exponential value of zero will gain an integer value every time that it is appears until it reaches a value of it having an exponential value of one integer less than what it had before the expansion occurred. If the two unknowns were added from to other, before the expansion, then the first parenthesis of the right hand side of the equation should have the same signs as were used on the other side of the equal sign and the second parenthesis should have a positive sign for the first term and this sign should be alternated with every new element that is added to the elements of the parenthesis. Also, if the two unknowns are subtracted from each other, before the expansion, then the first parenthesis of the right had side of the equation should have the same signs as were used on the other side of the equal sign and the second parenthesis should have a positive sign for all of terms in it.

$a^3 - b^3$ can be factored into $(a-b)(a^2 + ab^1 + b^2)$.

$a^4 - b^4 =$ $(a-b)(a^3 + a^2b + ab^3 + b^3)$

$a^5 - b^5 =$ $(a-b)(a^4 + a^3b + a^2b^2 + ab^3 + b^4)$

$a^3 + b^3$ can be factored into $(a + b)(a^2 - ab^1 + b^2)$.

$a^4 + b^4 =$ $(a + b)(a^3 - a^2b + ab^3 - b^3)$

$a^5 + a^5 =$ $(a + b)(a^4 - a^3b + a^2b^2 - ab^3 + b^4)$

Eg: $K^6 - B^6 = (K-b)(K^5 + K^4B + K^3B^2 + K^2B^3 + KB^4 + B^5)$

Pascal's Pyramid

$$1 \qquad (X+Y)^0 = 1$$

$$1 \quad 1 \qquad (X+Y)^1 = X + Y$$

$$1 \quad 2 \quad 1 \qquad (X+Y)^2 = X^2 + 2XY + Y^2$$

$$1 \quad 3 \quad 3 \quad 1 \qquad (X+Y)^3 = X^3 + 3X^2Y + 3XY^2 + Y^3$$

$$1 \quad 4 \quad 6 \quad 4 \quad 1 \qquad (X+Y)^4 = X^4 + 4X^3Y + 6X^2Y^2 + 4XY^3 + Y^4$$

$$1 \quad 5 \quad 10 \quad 10 \quad 5 \quad 1 \qquad (X+Y)^5 = X^5 + 5X^4Y + 0X^3Y^2 + 10X^2Y^3 + 5XY^4 + Y^5$$

$$1 \quad 6 \quad 15 \quad 20 \quad 15 \quad 6 \quad 1 \qquad (X+Y)^6 = X^6 + 6X^5Y + 15X^4Y^2 + 20X^3Y^3 + 5X^2Y^4 + 6X^1Y^5 + Y^6$$

$$1 \quad 7 \quad 21 \quad 35 \quad 35 \quad 21 \quad 7 \quad 1 \qquad (X+Y)^7 = X^7 + 7X^6Y + 21X^5Y^2 + 35X^4Y^3 + 35X^3Y^4 + 21X^2Y^5 + 7XY^6 + Y^7$$

The pyramid above allows one to find the coefficients, numbers that are not unknowns, of the various terms that are on the right hand side of this page.

Another way to find the coefficients is by using the one times table. First make a vertical and a horizontal line of ones. Add one to the first number this column. Record the result as the second number of the second column of numbers. Add that to the second number of the first column. Record the result as the third number of the second column of numbers. This can be done over and over again until you want to make the chart of numbers to have another column of numbers. To do that, repeat the process by adding one to the second number of the last column to get the second number of the third column of numbers. Then add that result to the third number of the previous column of numbers to get the third number of the third column of numbers.

1	1	1	1	1	1	1		2^0
1	2	3	4	5	6	7		2^1
1	3	6	10	15	21	28		2^2
1	4	10	20	35	56	84		2^3
1	5	15	35	70	126	210		2^4
1	6	21	56	126	252	462		2^5
1	7	28	84	210	462	924		2^6

Note that there are diagonal columns that run from any vertical column number one to a corresponding horizontal row number one. Each of these columns is either shaded or not shaded. The first sum of these is the number one which equals two to the zero power. The second sum of these is one plus one which equals two or two to the first power. The third of these is one plus two plus one which equals four or two to the second power. Jumping ahead, the seventh of these is the number one plus six plus fifteen plus twenty plus fifteen plus six and plus one which equals sixty-four or two to the sixth power. The far right column corresponds to the corresponding powers of two that are related to the sum of each shaded or non-shaded diagonal column of numbers. Again, this forms a pyramid when we look at it as being a slanted pyramid which I call the "Olaniyan Pyramid."

Some factoring examples

$25a^2 + 20a + 3$ This becomes $(5a + 3)(5a + 1)$.

Looking at a perfect square example we see

$5a^2 - 45b^2$ This becomes $(5)(a^2 - 9b^2)$ and then $(5)(a - 3b)(a + 3b)$.

Looking at a cube we see

$2a^3 - 16b^3$ This becomes $2(a^3 - 8b^3)$ and then $(2)(a^3 - [2b]^3) =$

$$2(a - 2b)(a^2 + a2b + [2b]^2).$$

$X^2Y^2 + 3Y^2 + X^2 + 3$ This can be factored as $(X^2Y^2 + 3Y^2) + (X^2 + 3)$.

We can factor out a Y^2.

$(y^2)(X^2 + 3) + (X^2 + 3) =$

$(X^2 + 3)(y^2 + 1)$

We can see that, here, factoring can show us patterns, such as there being two $(X^2 +3)$ factors.

__A Function__ .

A function is something that can be graphed. For a group of terms to be a function they must have only one Y value for each X value but there can be many X values for one y value.

Part Eight Problems

8A.) Solve for V when $(3V - 0.2V)^2 = 9$.

8B.) Solve for X when $[(3X - 2)^2 - (3X - 2) - 1]/(3X - 2)] = 20.95$.

8C.) Solve for X and Y. $Y - 3X = 0$.

8D.) A line starts at (4,0) and travels to (4, 100.00). Is this line a function?

8E.) Expand $(5K)^8 + (2N)^8$.

8F.) $(4B - B + 8)^2(B + 7) = $ What.

8G.) Factor $250a^2 - 640b^2$.

8H.) Factor $0.4a^2 - 81b$.

8I.) $25X^2 - 16 = $?

8J.) Factor $[(8a) - 6/b][64a^2 + 8a \cdot 6/b + 36/b^2]$.

8K.) Expand $(6a)^7 - c^7$.

8L.) A line starts at (0, 4) and travels horizontally to (166, 4). Is this line a function?

8M.) Expand $(2a)^2 + (N)^2$

8N.) Solve for X, $3X^2 - 9 = 66$.

8$\underline{O}$.) Expand $5K^8 + 2N^8$.

8P.) $(\sqrt{625})^2$ = What

\

Part Nine

More Problems

Do not use a calculator to do these questions

Questions

01.) On the graph of page three draw the function $Y = X^2 - X - 2$.

02.) $-66.08 + 43.23 - 4.09 + 1{,}008.19$ =

03.) $34 - 54 = 66 + X$, what is the value of X?

Football

And some Mathematics of Hidden Figures

In the movie "Hidden Figures" as the youngest student of her private school class, Katherine Johnson, had to solve for the below equation.

$$(X^2 +6X -7)(2X^2 - 5X -3) = 0$$

To solve this we can use the quadratic formula. This is a collection of coefficients. A coefficient is the number next to an unknown. To solve for a quadratic equation one should maneuver the numbers in such a manner that zero is on one side of the equals sign and the quadratic equation is used to find the value of the unknown/s.

In the above situation there are two sections, $(X^2 +6X -7)$ and $(2X^2 - 5X -3)$. While using the quadratic formula if either part, $(X^2 + 6X -7)$ or $(2X^2 - 5X -3)$, equals zero then the problem has been solved. The above equation has two sections and each section had three coefficients known as "a," "b," and "c." The "a" coefficients which are next to the X^2 term are one for the first set and two for the second set. The "b" coefficients which are next to the X^1 term are six for the first set and five for the second set, and the "c" coefficients which are next to, or multiplied by, the X^0 term is negative seven for the first set and negative three for the second set. Of course, X^0 equals one, 1. Here are the above terms again.

$$(aX^2 +bX^1 -cX^0)(aX^2 - bX^1 -cX^0) = 0 \quad \text{which is equal to}$$
$$(\ X^2 +6X^1 -7X^0)(2X^2 - 5X^1 -3X^0) = 0$$

We can divide the second part of this equation and the right side of this equation by two. This will reduce the coefficients of this part of the equation to no longer be a equals 2, b equals -5, and c equals -3 but to be one half of that previous value and become being 1 for a, being -2.5 for b, and being -3 for c. We can find the values of X by placing the above coefficients into the quadratic formula.

$$\frac{-b +/-(b^2 -4ac)^{0.5}}{2a} = 0$$

$-6 + [(6^2 - 4(1)(-7)]^{0.5}/2(1) = 0 \qquad -6 + (64)^{0.5}/2(1) = (1) \qquad$ AND

$-6 - [(6^2 - 4(1)(-7)]^{0.5}/2(1) = 0 \qquad -6 - (64)^{0.5}/2(1) = (-7).$

So the answer is (1, -7) which means X can be one and it can be negative seven.

Do not use a calculator to do these assigned problems.

Questions

04.) 456.98 is what percent of 200,200?

05.) Solve for X when the equation is $3X^2-2X + 12 = 16$.

06.) Change 3/57 into a percent and into a decimal number.

07.) Change 0.0123 into a percent and into a fraction.

08.) 92/46 – 622/34 =

09.) 459(3.3) – 400.309/9 =

10.) $456(5.05)^2 – X = 25.0500$, solve for X.

11.) 5/56 + 567/50 + 34/56 + 10/45 =

12.) -45/78 + 23X/39 + 452/7 -45/78 = 0, Solve for X.

13.) (-56/78)(78/1) =

14.) Answer with 8 decimal places $(234)(4/61)^3$ =

Questions (Do not use a calculator to do these assigned problems.)

15.) $(2/21)(23/51) =$

16.) $(6/201)(33/441) =$

17.) $X^2 - 12X - 15.75 = 6$. Solve for X.

18.) $2X(2-X^2) + 2X/(X)^3 = (2X)(2 - X^2) + (2X)(X^{-3})$ Solve for X.

19.) $20/65 - 40/130 =$

20.) $37X^4 89 - 25X/(2X)^{-3} = 52,688$

21.) Solve for X when the equation is $4X^2 - 20X + 5 = 10.2$

22.) Change 3.03 in to a percent.

23.) Solve for X when the equation is $2X^2 - 4X + 4 = 2$

24.) Solve for X when the equation is $4X^2 - 12X + 2 = -6$

25.) $452.00 - 809.05 =$

26.) $(2X - X^6)/X^{(8-2)}$ Simplify/reduce the terms.

27.) $5X - 2(X^2 - 4) = 4$ Solve for X.

28.) $22X[23/(4 + 16)] = 22$ Solve for X.

29.) $16X^2 - 9X + 4 = 3X$ Solve for X.

30.) Change 0.0123 into a percent and into a fraction.

Notes

Route One of Florida's Key's Areas, Plantation Key Colony, and Key West, Florida's Airport

Wilmington, Delaware

Part Ten.)
Pythagoras Theorem

Pythagoras came from Greece and studied, like others who arrived from Europe, in the present day area of Egypt. There, the Pythagoras Theorem was known not as the Pythagoras Theorem but as a mathematical fact that was known before he arrived to Kemit, ancient Egypt.

This theorem covers many calculations that can be represented on the X, Y, Z coordinate system. First, this theorem covers simple calculations such as <u>find the length between 2X and 3X</u>. The answer 1X is used in most college algebra or geometry courses. Here, it includes only one dimension of our three dimensional world. Also, often we will not print the unknown, X or Y or Z, etc., so the distance between 2X and 3X will be printed as the distance between 2 and 3. Note: distance is always positive in value. Second, this theorem often is use to find the distance between locations such as (2X, 4Y) and (-2X, 6Y). Here, it includes two dimensions, X and Y, of our three dimensional world. Yet, third, I like to take this theorem to "another level," which is the "Real World," that is three dimensional and includes the Z axis. Here, the formula is

$$[X^2 + Y^2 + Z^2]^{0.5}$$

Note that the 0.5 power is the same as getting the square root of a number. (X or any number or thing)^2 and then raised to the 0.5 power will equal X or that same anything, except if we start with a number like -5, a negative five, because raising -5 to the second power gives us twenty-five and getting the square root of twenty-five, or raising twenty-five to the one-half power, will result in a positive five, not a negative five.

First and second, again, this can be used to find the distance between two spots of a floor. Third, again, this can be used to find the distance between a spot at the middle of the floor and the place where two walls of a room meet at the ceiling of the room. This last use of the formula is used in biology, architecture, physics, but it is not used in economics. (Interesting)

The part of the theorem that is taught in high school and many college courses includes the XY coordinate system that is shown below. Each intersection of two lines can be represented with an X coordinate and a Y coordinate.

For example, in the graph, the top intersection of lines is known as (14X, 11Y) but it is usually represented as (14, 11) or 14,11. In this graph, the lines of positive X values are left of the printed numbers of the graph, the lines of negative X values are right of printed numbers of the graph, the lines of positive Y values are below the printed numbers of the graph, the lines of negative Y values are above the printed numbers of the graph, and the origin is not labeled as it usually is labeled, 0,0.

To find the distance between the -3,-7 and the 1,6 locations the $[X^2 + Y^2 + Z^2]^{0.5}$ formula will be used. Here, we have X as being the change in X values. This is found by subtracting one X value from the other X value before this sum is raised to the second power. $(-3-1)^2$

Y is the change in Y values and when this change is raised to the second power it is $(-7-6)^2$. Also Z is the change in Z values and it is then raised to the second power, $(0 - 0)^2$. This "boils" down to being $[(-4)^2 + (-13)^2 + (0)^2]^{0.5}$. Next, $[(-4)^2+(-13)^2 + (0)^2]^{0.5} = (16 +169)^{0.5} = 13.60147051$ or 13.6015 if we only use four decimal places in this answer.

What happens if the graphed object does not only have the shown two dimensions but has a third dimension, Z. In this question we have the original X and Y coordinates with a Z coordinate of -15, -3, -7, -15, as one point of this three dimensional figure and we have the other original X and Y coordinates with a Z coordinate of -5 of this three dimensional figure. So, 1, 6, -5 is the other point of this three dimensional figure. The distance between these two points, -3, -7, -15 and 1, 6, -5, can be found from the above formula, $[X^2 + Y^2 + Z^2]^{0.5}$.
$[(-3-1)^2+(-7-6)^2+(-15- -5)^2]^{0.5}=[(-4)^2+(-13)^2+(-10)^2]^{0.5}$
$= [(16 + 169 + 100]^{0.5} = [(285)]^{0.5} = 16.8819$, if we, again, use only four decimal places.

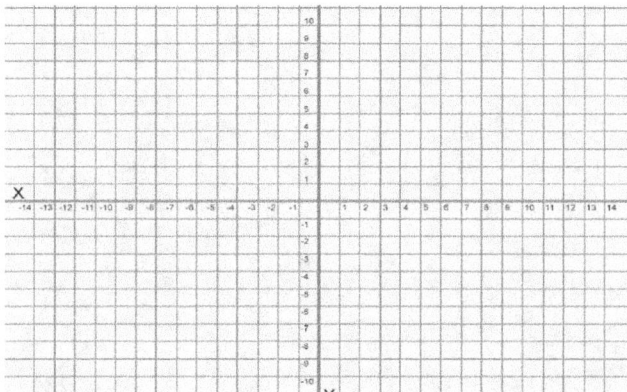

Again, this theorem covers simple calculations such as finding the distance between 2X and 3X. This is usually expressed as two minus three, 2-3. By using the Pythagoras formula the answer can be found.

$$[X^2 + Y^2 + Z^2]^{0.5} =$$
$$[(2\text{-}3)^2 + (0\text{-}0)^2 + (0\text{-}0)^2]^{0.5} =$$
$$[(\text{-}1)^2 + (0)^2 + (0)^2]^{0.5} =$$
$$[(\text{-}1)^2]^{0.5} = [\ 1\ ]^{0.5} = 1$$

Pythagoras Theorem with Three Dimensions

(The scales of the pictures are not necessarily correct.)

With figure one, our eyes may try to fool us by telling us that point E is at the front left side of the ceiling of the figure even though it really is at the back left side of the floor below Point G. Figure two shows a better picture. Here, you are, without seeing, to believe that there is a point E that is equal distance from point G just like Point H is from Point F.

Let's notice this box's size. The size of the distance between Points A and B is the same as the distance between Points B and F, three and one half feet, 3.5ft. The distance between Points F and H is 3.391164992. The picture has a X range, left to right, from Point A to B, Point C to D, Point E to F, and Point G to H. The box has a Y range, forward and back, from Point A to E, B to F, C to G, and D to H. This box has a Z range, floor to ceiling, from A to C, B to D, E to G, and F to H.

What is the greatest distance between the points of this box?

This may be seen as a tricky question. To solve it one can just look at the room that she or he is in at the present time. Here, one can see that the distance between corners is greatest if one finds two corners that do not share the same walls and one of the two is on the floor and the other is at the ceiling. In a three dimensional figure, with parallel plains, there are four lines of this nature. Below, one of them from the figure is shown.

$$[(A\text{-}b)^2 + (B\text{-}F)^2 + (F\text{-}H)^2]^{0.5} =$$

$$[(3.5)^2 + (3.5)^2 + (3.391164992)^2]^{0.5}$$

$$[(12.25)+(12.25)+(11.5)]^{0.5} = [36)]^{0.5} = 6$$

Figure One

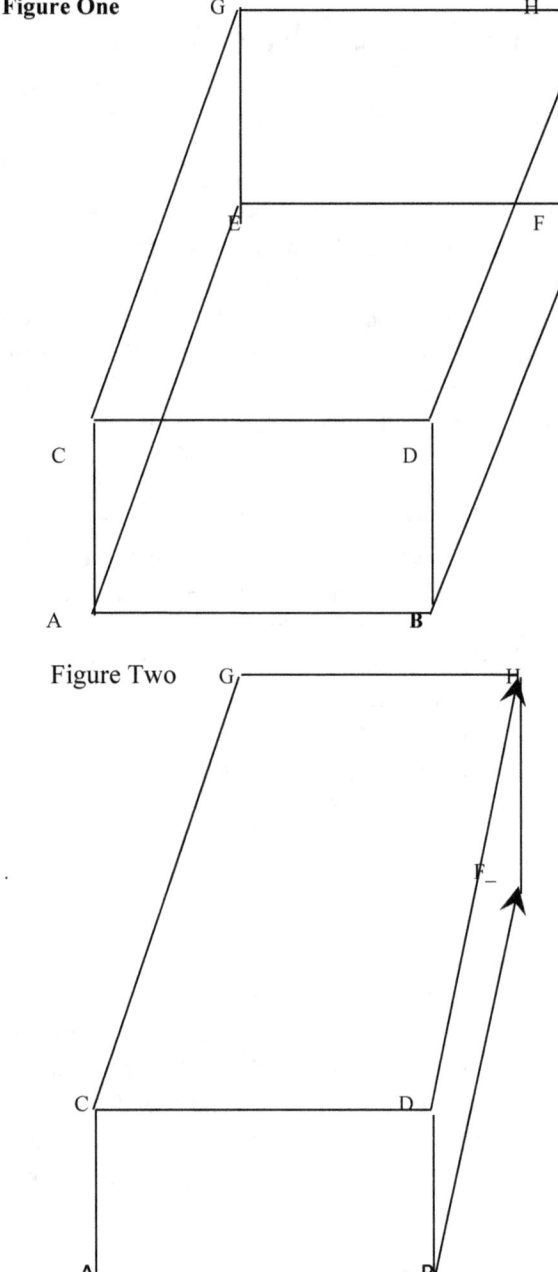

Figure Two

Area, etc.

The area of a rectangle is found by multiplying the X coordinate size by the Y coordinate size. Here, a square is a rectangle that has four equaled sizes and ninety degree angles. The area of a cube is found by multiplying the X coordinate side by the Y coordinate size and then by the Z coordinate size. So, if a box has sides that are all 34.5 inches the area of any side is 119.025 inches and the volume of the box is 41,063.625 cubic inches.

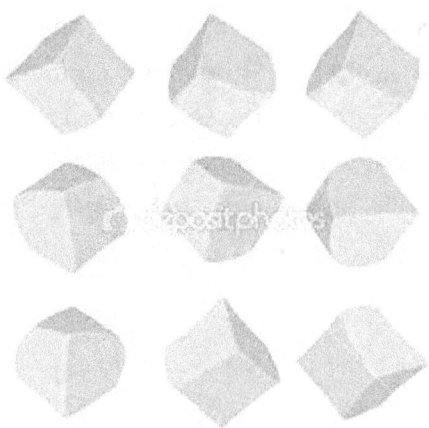

The area of a circle is pi r^2 or pi diameter^2 divided by four. Here pi is valued at 3.141592654 but is often limited to the 3.14 value. Every circle has a diameter and half of the diameter is the radius. In the above equation r is known as being the radius. The circumference of a circle is 2 pi radius or pi times the circle's diameter. The volume of a sphere is (4/3)(pi radius^3) or pi times the diameter cubed before being divided by six.

Finding the distance between two points is very similar to working with a triangle. Of course triangles bring into our minds the subject of triangles or the study of three angles that is also known as trigonometry.

Going further, let the horizontal line of the triangle be the X axis and let the vertical line be the Y axis. If X is 4 yards long and Y is 3 yard long then H, or the hypotenuse, is $(4^2 + 3^2)^{0.5}$ long. So, the hypotenuse line tells us the distance between two points. Should we create an infinite amount of hypotenuse lines that all point away from a central point and that all have the same length then if we connect the end points of each of these lines, that is far from the central point of where these lines all started, here, we will have created a circle. Should we connect a horizontal and a vertical line from the center of the circle to the circumference of the circle we will be able to notice a few patters. A diameter line has the length of a circle and half of this line is called the radius which ranges from the center of the circle to the circle's circumference. So, again, from the center of the circle to anywhere on the circumference of the circle is the length of the hypotenuse line, H.

From here we can find that $X^2 + Y^2 = H^2$ because $H = (X^2 + Y^2)^{0.5}$

This also means that $X^2 = H^2 - Y^2$ or $X = (H^2 - Y^2)^{0.5}$

$Y^2 = H^2 - X^2$ or $Y = (H^2 - X^2)^{0.5}$. So, by knowing the value of any of one of the variables, X and Y, the other variable can be found. Note that H is not a variable. It is a constant, meaning that unlike variables its value cannot be changed.

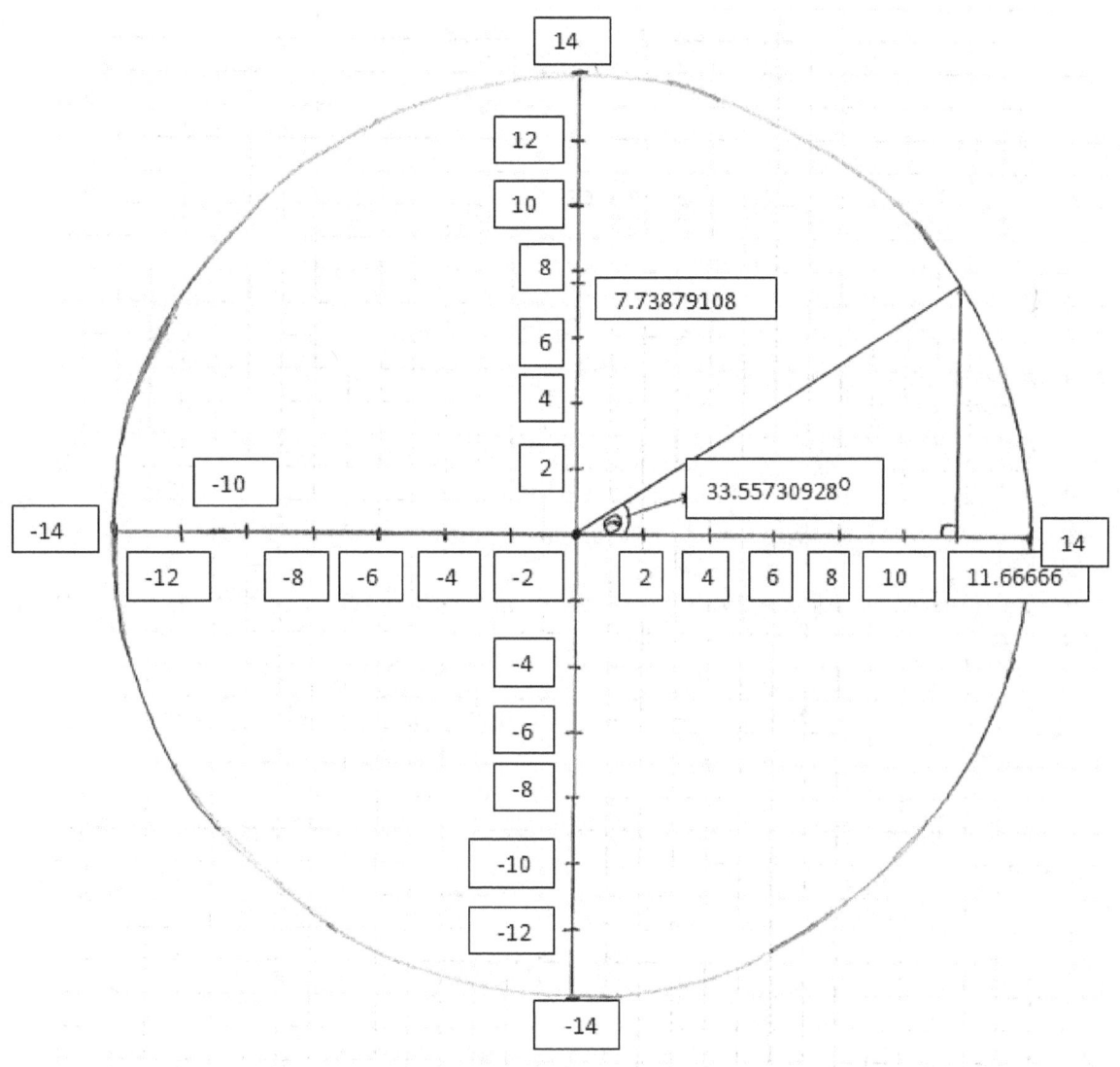

Questions Do not use a calculator to do these assigned problems.

31.) $-2X - 44 = 22$, what is the value of X?

32.) What is the distance between points (-25X, 204Y, 0Z) and (14X, 2Y, 13.98Z)?

33) $(X^{0.3})^{20.2}X^2 =$ What?

34.) **Kwanzaa Homework:** **What is** **Dr. Dean's relationship to the computer?**

Part Ten Problems

(Do all multiplication, division, addition,
and subtraction without use of a calculator.)

10A.) Multiply -10.8^2 by -1.50^{-1}.

10B.) What is the volume of a box that has an area of 0.40 feet and a 2.2 area?

10C.) A rectangle has four sides of 16 inches. How many square feet are in it?

10D.) Using six decimal places, a circle has an area of 25 feet and a diameter of 5.64895 inches. What is this circle's radius?

10D.) One circle with a radius of 4 inches and another circle with a radius of 16.5 inches are situated on the X, Y, Z coordinate system, respectively, at 3, -4, -2 and 4.5, 202, -8.1. What is the distance between these two circles?

10E.) How many cubes that have a twenty inch side can be placed into a plane that is 8.5 feet by 90.5 feet?

10F.) $4^{16} / (64^2$ multiplied by $64/8.080^2)$ after expressing the answer in four decimal places is what?

10G.) The distance between two circles that both have a radius of 22 feet and both have a center that is situated at X equals 2.6 but one is located is located at Y equals 2.5 while the other is located at Y equals 5.59.

10H.) Express $725.0725/100^2$ as a percent.

10I.) What percentage number represents ten thousand^2?

10J.) Divide 00 by 1/1. What is the answer?

10K.) A circle with a diameter of 14 inches has an X value of 6.55 at its circumference. What is the corresponding Y value of this circle?

10L.) $0.20^3 / 64 =$ what?

10M.) Multiply 0.8 by 15^3.

10N.) Divide thirty-eights by two hundred and forty-one.

10O.) Divide -0.48 by -16.

10P.) Raise 10.03 to the 2.02 power.

10Q.) Factor $750d^2 + 187.5a^2$.

10R.) Factor $250a^2 - 640b^2$.

10S.) Factor $0.4a^2 - 81b$.

10T.) $1.1001/(10,022.40/1.1001^3)= ?$.

10U.) What percentage number represents one-halve cubed?

10V.) Write 0 percent as a percentage number.

The End of Part Ten and the start of Part Eleven

Part Eleven-Equations

Walt Mossberg

The Steve Jobs I Knew

October 5, 2011 at 6:50 pm PT

Asa Mathat | AllThingsD.com

That Steve Jobs was a genius, a giant influence on multiple industries and billions of lives, has been written many times since he retired as Apple's CEO in August. He was a historical figure on the scale of a Thomas Edison or a Henry Ford, and set the mold for many other corporate leaders in many other industries.

He did what a CEO should: Hired and inspired great people; managed for the long term, not the quarter or the short-term stock price; made big bets and took big risks. He insisted on the highest product quality and on building things to delight and empower actual users, not intermediaries like corporate IT directors or wireless carriers. And he could sell. Man, he could sell.

As he liked to say, he lived at the intersection of technology and liberal arts.

But there was a more personal side of Steve Jobs, of course, and I was fortunate enough to see a bit of it, because I spent hours in conversation with him, over the 14 years he ran Apple. Since I am a product reviewer, and not a news reporter charged with covering the company's business, he felt a bit more comfortable talking to me about things he might not have said to most other journalists.

Even in his death, I won't violate the privacy of those conversations. But here are a few stories that illustrate the man as I knew him.

61

The Phone Calls   I never knew Steve when he was first at Apple. I wasn't covering technology then. And I only met him once, briefly, between his stints at the company. But, within days of his return, in 1997, he began calling my house, on Sunday nights, for four or five straight weekends. As a veteran reporter, I understood that part of this was an attempt to flatter me, to get me on the side of a teetering company whose products I had once recommended, but had, more recently, advised readers to avoid.

Yet there was more to the calls than that. They turned into marathon, 90-minute, wide-ranging, off-the-record discussions that revealed to me the stunning breadth of the man. One minute he'd be talking about sweeping ideas for the digital revolution. The next about why Apple's current products were awful, and how a color, or angle, or curve, or icon was embarrassing.

After the second such call, my wife became annoyed at the intrusion he was making in our weekend. I didn't.

Later, he'd sometimes call to complain about some reviews, or parts of reviews — though, in truth, I felt very comfortable recommending most of his products for the average, non-techie consumers at whom I aim my columns. (That may have been because they were his target, too.) I knew he would be complaining because he'd start every call by saying "Hi, Walt. I'm not calling to complain about today's column, but I have some comments, if that's okay." I usually disagreed with his comments, but that was okay, too.

The Product Unveilings   Sometimes, not always, he'd invite me in to see certain big products before he unveiled them to the world. He may have done the same with other journalists. We'd meet in a giant boardroom, with just a few of his aides present, and he'd insist — even in private — on covering the new gadgets with cloths and then uncovering them like the showman he was, a gleam in his eye and passion in his voice. We'd then often sit down for a long, long discussion of the present, the future, and general industry gossip.

I still remember the day he showed me the first iPod. I was amazed that a computer company would branch off into music players, but he explained, without giving any specifics away, that he saw Apple as a digital products company, not a computer company. It was the same with the iPhone, the iTunes music store, and later the iPad, which he asked me to his home to see, because he was too ill at the time to go to the office.

The Slides   To my knowledge, the only tech conference Steve Jobs regularly appeared at, the only event he didn't somehow control, was our D: All Things Digital conference, where he appeared repeatedly for unrehearsed, onstage interviews. We had one rule that really bothered him: We never allowed slides, which were his main presentation tool.

One year, about an hour before his appearance, I was informed that he was backstage preparing dozens of slides, even though I had reminded him a week earlier of the no-slides policy. I asked two of his top aides to tell him he couldn't use the slides, but they each said they couldn't do it, that I had to. So, I went backstage and told him the slides were out. Famously prickly, he could have stormed out, refused to go on. And he did try to argue with me. But, when I insisted, he just said "Okay." And he went on stage without them, and was, as usual, the audience's favorite speaker.

Ice Water in Hell

For our fifth D conference, both Steve and his longtime rival, the brilliant Bill Gates, surprisingly agreed to a joint appearance, their first extended onstage joint interview ever. But it almost got derailed.

Earlier in the day, before Gates arrived, I did a solo onstage interview with Jobs, and asked him what it was like to be a major Windows developer, since Apple's iTunes program was by then installed on hundreds of millions of Windows PCs.

He quipped: "It's like giving a glass of ice water to someone in Hell." When Gates later arrived and heard about the comment, he was, naturally, enraged, because my partner Kara Swisher and I had assured both men that we hoped to keep the joint session on a high plane.

In a pre-interview meeting, Gates said to Jobs: "So I guess I'm the representative from Hell." Jobs merely handed Gates a cold bottle of water he was carrying. The tension was broken, and the interview was a triumph, with both men acting like statesmen. When it was over, the audience rose in a standing ovation, some of them in tears.

The Optimist

I have no way of knowing how Steve talked to his team during Apple's darkest days in 1997 and 1998, when the company was on the brink and he was forced to turn to archrival Microsoft for a rescue. He certainly had a nasty, mercurial side to him, and I expect that, then and later, it emerged inside the company and in dealings with partners and vendors, who tell believable stories about how hard he was to deal with.

But I can honestly say that, in my many conversations with him, the dominant tone he struck was optimism and certainty, both for Apple and for the digital revolution as a whole. Even when he was telling me about his struggles to get the music industry to let him sell digital songs, or griping about competitors, at least in my presence, his tone was always marked by patience and a long-term view. This may have been for my benefit, knowing that I was a journalist, but it was striking nonetheless.

At times in our conversations, when I would criticize the decisions of record labels or phone carriers, he'd surprise me by forcefully disagreeing, explaining how the world looked from their point of view, how hard their jobs were in a time of digital disruption, and how they would come around.

This quality was on display when Apple opened its first retail store. It happened to be in the Washington, D.C., suburbs, near my home. He conducted a press tour for journalists, as proud of the store as a father is of his first child. I commented that, surely, there'd only be a few stores, and asked what Apple knew about retailing.

He looked at me like I was crazy, said there'd be many, many stores, and that the company had spent a year tweaking the layout of the stores, using a mockup at a secret location. I teased him by asking if he, personally, despite his hard duties as CEO, had approved tiny details like the translucency of the glass and the color of the wood.

He said he had, of course.

The Walk After his liver transplant, while he was recuperating at home in Palo Alto, California, Steve invited me over to catch up on industry events that had transpired during his illness. It turned into a three-hour visit, punctuated by a walk to a nearby park that he insisted we take, despite my nervousness about his frail condition.

He explained that he walked each day, and that each day he set a farther goal for himself, and that, today, the neighborhood park was his goal. As we were walking and talking, he suddenly stopped, not looking well. I begged him to return to the house, noting that I didn't know CPR and could visualize the headline: "Helpless Reporter Lets Steve Jobs Die on the Sidewalk."

But he laughed, and refused, and, after a pause, kept heading for the park. We sat on a bench there, talking about life, our families, and our respective illnesses (I had had a heart attack some years earlier). He lectured me about staying healthy. And then we walked back.

Steve Jobs didn't die that day, to my everlasting relief. But now he really is gone, much too young, and it is the world's loss.

▇C The Green House Effect

One of our main physical elements is carbon. Every section of your body has it. Many, human and non-human, have died and became buried underground. Much later the carbon from their buried bodies was used to fuel automobiles and machinery. This helped some humans to create the "Age of Enlightenment." With this so-called enlightenment some humans were fueled to conquer the lands of other humans. Many of the human victimizers made more advancements as results of the wars they launched on the so-called "developing" or "underdeveloped" nations. This advancement created lands with much modern developments and not very much green lands per a resident of the lands. Many of the human victims, especially those of other nations, did not make as much advancements and have lands that have much greenery per a resident.

After years of creating carbon abuse, humans have decided to reduce the emissions of carbon dioxide that creates a warm environment called "the Green House Effect." The balance of trade records and other economics elements have not been solutions to this pollution problem, that creates great wealth for some nations as it harms many less developed nations.

Much of the bad balance of trade is an issue that has not entered the global warming meetings and the 2009 Climate Conference of various nations. Many so-called underdeveloped nations have large green lands. Their trees, etc., create oxygen, a solution to pollution. These nations, including Brazil, The Democratic Republic of the Congo, India, and others, should be compensated by the considered to be first world nations, which include the United States of America, Japan, England, France, Italy, Germany, and Russia among others. This is not necessarily reparations yet it is related to respecting the balance of trade, even though the production and delivery of oxygen is not respected as a product in the world's various recorded balances of trade. So-called third-world nations have been negligently giving to others, the leading top seven or so economic leaders included, without receiving any recorded consideration, money, for too many years. It is time for these nations to be paid for their development of oxygen which helps the world exists and not be destroyed by the Green House Effect.

A Function has one corresponding Y value for every X value.

Two positive numbers, or unknowns, multiplied or divided together result in a positive number.

Two negative numbers, or unknowns, multiplied or divided together result in a positive number.

A positive and a negative number multiplied or divided together result in a negative number.

All numbers have an exponent and when the exponent is one the number one is rarely shown.

When multiplying two equal numbers add their exponents, example $X^8 X^{56} = X^{64}$. Also $X^{0.5}$ is the square root of X.

When raising a number to a second exponent other than its original exponent quantity, multiply the two exponents to create a new exponent. Here is an example; $(X^8)^{56} = X^{448}$. Another example is $[(3)^4]^5$. It is the same as $(3)^4$ $(3)^4$ $(3)^4$ $(3)^4$ $(3)^4$ which equals $(3)^{20}$.

The plot of an equation with one unknown has a form of mx-b, where m is the slope of the function or the "rise" of the function divided by the change, or "run," in X of the function. The Y intercept is b. When X is zero the function intersects the Y axis at b. To solve equations of one unknown set it so that unknowns are on one side and coefficients are on the other side of the equation. Below is an example of how this is done.

To graph the equation 25X we should note that the slope is 25 and the b term is not shown because it is zero. The result will be a straight line that increases with twenty-five Y values every time the X value increases by one. A perpendicular line to it will have a slope of -25 instead of 25.

Using slopes the formula Sarafu was created.

Find ten coins, made up only of U.S.A. quarters, dimes, nickels, and pennies, that sum to seventy-eight cents.

This method, Sarafu – which is Swahili for the English words Coin/coins -of solving these types of problems was designed by Olaniyan Adefumi.

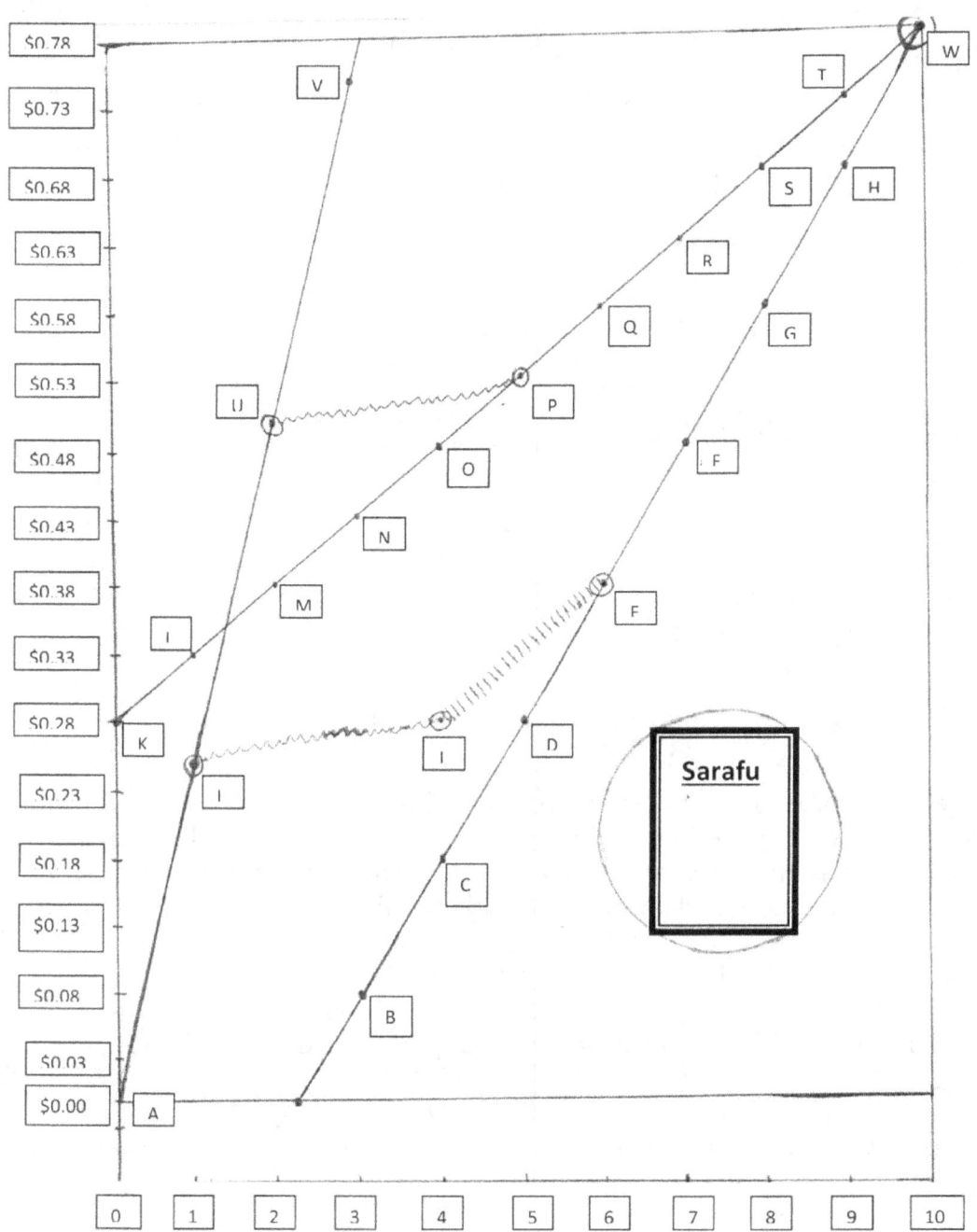

On the graph we can travel from the Origin Point to the last point which is Point W. From the Origin Point to Point I include just one quarter. From Point I to Point J, which is done on a penny line, there are three pennies counted. From Point J to Point E we travel on a Nickel Line and count two nickels. Last From Point E to Point W we count four dimes. The sum of one quarter, three pennies, two nickels, and four dimes adds to a value of seventy-eight cents.

Re: Sarafu
...
     "Torrence, Bruce" <btorrenc@rmc.edu>
From: ...
     Add to Contacts
To: Olaniyan Adefumi <oadefumi@yahoo.com>

Here's the email I sent earlier:
Hi Olaniyan,

In Mathematica, one can simply type:
IntegerPartitions[119, {13}, {1, 5, 10, 25}]

And the output will be:
{{25, 25, 25, 10, 10, 5, 5, 5, 5, 1, 1, 1, 1}, {25, 25, 10, 10, 10, 10, 10, 10, 5, 1, 1, 1, 1}}

So there are two ways to make change for $1.19 using 13 coins (without half-dollars). If one replaces "{13}" by "119", each of the 343 ways of making change for $1.19 will be output. Simply adding "50" to the list in the last argument allows for including fifty-cent coins. The theory of generating functions, as it pertains to partitions, is useful here. The following link is a good reference:
http://www.math.upenn.edu/%7Ewilf/PIMS/PIMSLectures.pdf

The fact that there is a Mathematica command to do this implies the existence of a fast algorithm.

Your Sarafu method is very clever, and provides a nice visualization tool, but this is well-trodden mathematical ground.

Best regards,
- Bruce Torrence

Questions

35.) $-2X - 44 = 22$ What is the value of X

36.) $(-234.009)(433.09) =$ What?

37.) $\dfrac{3B^3W^2 - 12B^6W^2}{2B} =$ _____ when B = 1.4 and W = -2.0.

Sarafu

You have one hundred dollars to purchase one hundred animals. Pigs cost $3.00, horses cost $10.00, and chickens cost $0.50. How many of each kind of animal can you purchase if you spend all of your funds of one hundred dollars, $100.00?

The process named Sarafu was created years ago to tell how many quarters, dimes, half-dollars, nickels, and pennies one must have to have a particular amount of money if the number of coins is limited to a certain amount.

In the below graph, by traveling along the horses' line to where it meets the chicken's line and following that line to where it meets the pig's line and follow that line until the point that represents one hundred dollars and one hundred animals is reached, at the far right top section of the graph, you will see that the first intersection represents five horses, the second intersection represents ninety-four chickens, and the third intersection represents one pig.

THIS SUM IS ONE – HUNDRED ANIMALS.

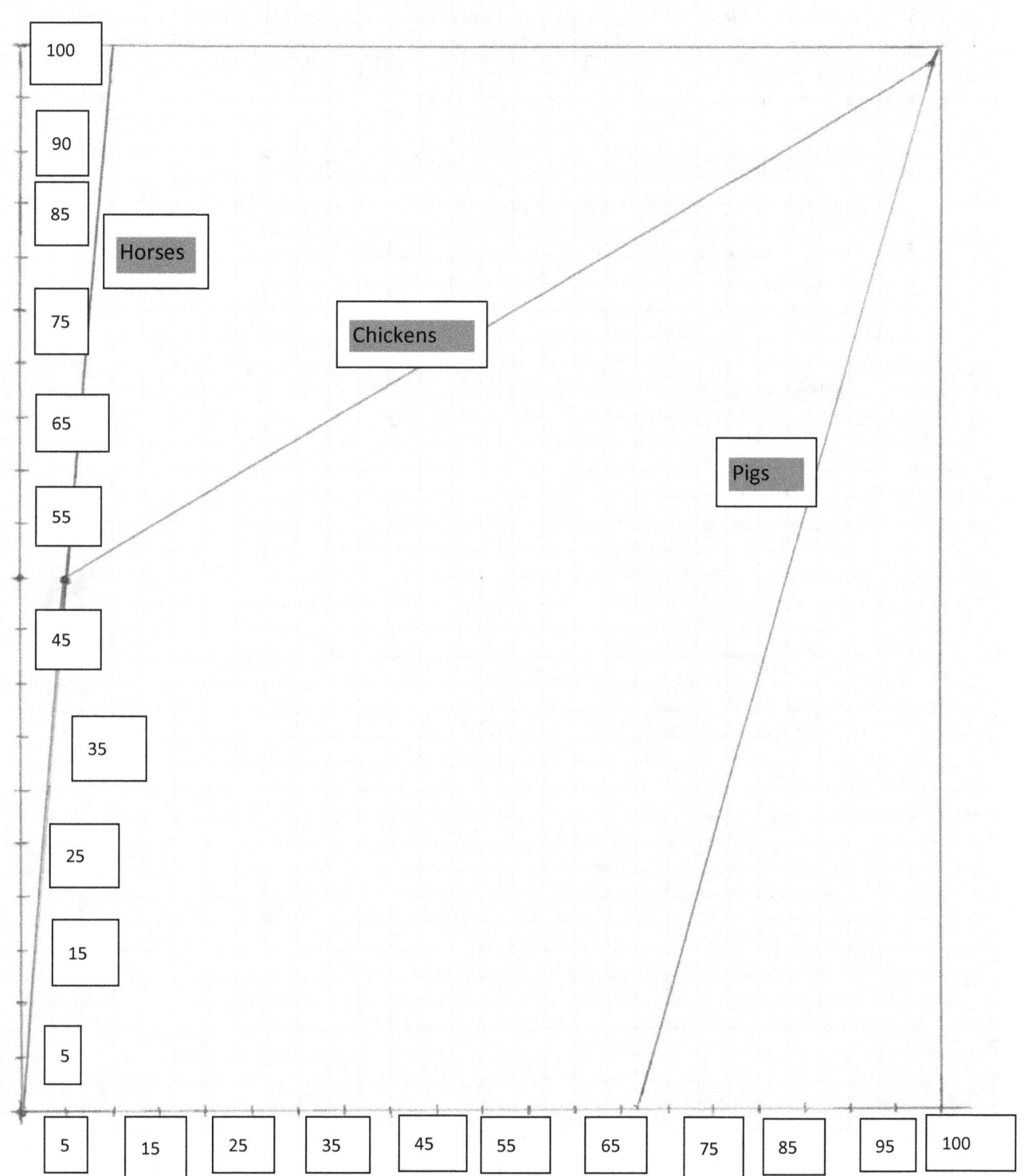

By traveling along the horses' line to where it meets the chicken' line and following that line to where it meets the pig's line and follow that line until the point that represents one hundred dollars and one hundred animals is reached, at the far right top section of the graph, you will see that

the first intersection represents   5 horses,
the second intersection represents   94 chickens, and
the third intersection represents   1 pig.

Sarafu

5	Horses	$10.00 each		$ 50.00
94	Chicken cost	$ 0.50 each		$ 47.00
1	Pigs cost	$ 3.00 each		$ 3.00
100	Total			$100.00

"Don't believe the Hype" by Chuck D

Notes

Some things about **Equations**

    Look carefully at the spelling of this word. Related to it is the word "equal." Equations, thus, must have an equal sign and two things that are equal to each other, by definition. With this definition are the "unknowns" and the "knowns." The unknowns are also called "variables." Variables are often represented by the use of letters. Any letter or set of letters can be used. Some of the usually used letters are X, x, Y, y, Z, z, H, h, I, I, J, j, K, k, A, a, B, b, C, and c. The "knowns" are also called "coefficients." They are usually numbers that are well specified.

    To solve an equation is to find the value of the "unknowns." This can be done by adding a quantity to the equation, subtracting a quantity from the equation, dividing the equation by a quantity, multiplying the equation by a quantity, or raising the equation to a particular power. To make this look simple there are some examples.

    A.) $3X - 14 = -2$

        To solve this we first should place all of the unknowns on one side of the equal sign and the coefficients on the other side of the equation. To do this a fourteen, 14, will be combined to both sides of the equation.

        $3X - 14 + 14 = -2 + 14$ Next, the numbers will be summed.

        $3X = 12$ Next, to get the coefficients to be on the opposite side of the equation from the unknowns the entire   equation will be divided by three, 3.

        $3X/3 = 12/3$

        $X = 4$ Here, we see that X is equal to four.

    This is how solving equations is often done when there is one unknown, such as X.

    What happens where there is more than one unknown? Here, we have to have more equations. For every unknown we must have an equation, in order for us to solve for all of the unknowns. To solve these equations, called matrices, the method used is often called the "Triangular Method." It forms the shape of a triangle when it is used. Let us solve for this small matrix of equations. $-4X - 5Y = -6$
        $3X - 2Y = 10$

To begin we should added values to both equations so that the results will be set in a manner that the right hand side of the equations equal zero. To get this result we can add a six to the first equation and we can add a negative ten to the second equation.

        $-4X - 5Y + 6 = -6 + 6$
        $3X - 2Y - 10 = 10 - 10$
        After summing the results we will have

$-4X - 5Y + 6 = 0$ and $3X - 2Y - 10 = 0$

To solve for these two equations we must multiply the first equation by a factor that will change the coefficient of the first term so that when the result of this multiplication is added to the second, or next, equation the result will be equal to zero.

When the first equation is multiplied by seventy-five percent, 0.75, and added to the second equation, the result is what is printed below.

$$-4X - 5Y + 6.0 = 0$$
$$- 5.75Y - 5.5 = 0$$

To solve for Y we can add five and one-half to both sides of the second equation.

$$-4X - 5.00Y + 6.0 = 0$$
$$-5.75Y - 5.5 + 5.5 = 0 + 5.5$$

This will be the same as

$$-4X - 5.00Y + 6 = 0$$
$$- 5.75Y = 5.5$$

Next, divide the second equation by negative five and three-fourths, -5.75, where to divide the equation means to do it on both sides of the equation. The results are

$$-4X - 5Y + 6 = 0$$

$$Y = -0.956521739$$

Next, we want to solve for X. To do this we go to the first equation, -4X – 5Y, and replace Y with the newly found value of -0.956521739. The result is

$$-4X - 5(-0.956521739) + 6 = 0 \quad \text{Which is the same as}$$

$$-4X + 4.782608695 + 6 = 0 \text{ or.}$$
$$-4X + 10.782608695 = 0$$

To solve for X we must add -10.782608695 to the equation.

$$-4X + 10.782608695 - 10.782608695 = 0 - 10.782608695.$$

Going further will give the result that is negative four, -4, times the value of X.
$$-4X = -10.782608695$$
So, dividing this equation by negative four, - 4, will show us the value of X. $-4X/-4 = -10.8926998695/-4$

 IN CONCLUSION   X = 2.695652173

<u>Last, going back to the original equations and replace X and Y with their newly found values.</u>

-4X - 5Y = -6

3X - 2Y = 10

Which is the same as

$$-4(2.695652174) - 5(-0.956521739) = -6$$
$$3(2.695652174) - 2(-0.956521739) = 10$$

_____

Homework: -5X + 10 = Y

3X -   0 = Y

The Homework Solution:

Just like with the other matrix problems we need to have each equation equal to zero. To do this, we can subtract Y from each of the above equations.

$$-5X + 10 - Y = Y - Y$$

$$3X - 0 - Y = Y - Y$$

The beginning two equations are

-5X – Y + 10 = 0   and   3X – Y –   0 = 0.

Multiply the first equation by 3/5 and add the result to the second equation.

(-5X – Y + 10)(3/5) = 0(3/5)   The results will be

[(-5X/1)(3/5) -3Y/5 + 30/5] = (0)(3/5) Which is the same as

-3X -3Y/5 + 6 = 0   and the second equation, once added to -3X -3Y/5 + 6, will become

-8Y/5   + 6 = 0

We can solve for Y by adding a negative six, -6, to the equation.

-8Y/5 + 6 -6 = 0 -6   This is the same as   -8Y/5   = -6

Next, multiply the equation by five, 5, and divide it by a negative eight, -8.

(-8Y/5)(5)/-8   =   (-6)(5)/(-8) The result will be Y on the left hand side of the equation and on the right had side of the equation there will be a negative thirty, -30, divided by a negative eight, -8, which will give a result of three and three-fourths, 3.75. This is the same as   Y = 3.75.

We can choose any of the two beginning equations to solve for X.   I will work with the first equation.

$$-5X - Y + 10 = 0 \qquad \text{or} \qquad -5X - 3.75 + 10 = 0 \text{ or } -5X + 6.25 = 0$$

After adding a negative six and one quarter to this equation only a negative five X will remain on the left hand side of the equation while on the right hand side of the equation there will be a negative six and one quarter, -6.25.

$$-5X + 6.25 - 6.25 = 0 - 6.25$$

$-5X$ $\qquad\qquad\qquad = \qquad - 6.25$ Next, to find the value of X , divide the equation by negative five, -5. The result is that X is equal to 1.25.

The second equation will give the same results.
$3X \qquad - Y \qquad - 0 = 0$

$3(1.25) \quad - 3.75 \quad - 0 = 0$

Don't worry, Life is much more complicated.

Counting atoms Avogadro's number Amadeo Avogadro (1766-1856) never knew his own number; it was named in his honor by a French scientist in 1909. its value was first estimated by Josef Loschmidt, an Austrian chemistry teacher, in 1895.

Owing to their tiny size, atoms and molecules cannot be counted by direct observation. There are, however, a number of indirect methods that enable us to estimate the number of these particles in a sample of an element or compound. Once this has been done, we know the number of formula units (to use the most general term for any combination of atoms we wish to define) in any arbitrary weight of the substance. The number will of course depend both on the formula of the substance and on the weight of the sample. But if we consider a weight of substance *that is the same as its formula (molecular) weight expressed in grams*, we have only one number to know: ***Avogadro's number***, $6.022141527 \times 10^{23}$, usually designated by N_A.

What is the special significance of this huge number, 6.02×10^{23} ? To help you understand this extremely important point, take a moment to convince yourself of the reasoning embodied in the following sequence of problems:

?? The atomic weights of oxygen and of carbon are 16.0 and 12.0, respectively. How much heavier is the oxygen atom in relation to carbon?

Solution: Atomic weights represent the relative masses of different kinds of atoms. This means that the atom of oxygen has a mass that is $16/12 =$ **4/3** ≈ 1.33 as great as the mass of a carbon atom.

?? The absolute mass of a carbon atom is 12.0 unified atomic mass units (What are these?). How many grams will a single oxygen atom weigh?

Solution: The absolute mass of the carbon atom is 12.0 **u**, or $12 \times 1.6605 \times 10^{-27}$ g
$= 19.9 \times 10^{-27}$ kg. The mass of the oxygen atom will be 4/3 greater, or **2.66×10^{-26} kg**.

Alternatively: $(12 \text{ g/mol}) \div (6.022 \times 10^{23} \text{ mol}^{-1}) \times (4/3) =$ **2.66×10^{-23} g**.

?? Suppose that we have *N* carbon atoms, where *N* is a number large enough to give us a pile of carbon atoms whose mass is 12.0 grams. How much would the same number, *N*, of oxygen atoms weigh?

Solution: The collection of *N* oxygen atoms would have a mass of $4/3 \times 12$ g = **16.0 g**.

So what value must N have in order to make the weight of a pile of N atoms of any kind numerically equal to the atomic weight of the element? **The answer is just Avogadro's number $N_\mathrm{A} = 6.022141527 \times 10^{23}$.**

Amedeo Avogadro Biography

History of Avogadro

By <u>Anne Marie Helmenstine, Ph.D.</u>, About.com Guide

Amedeo Avogadro was born August 9, 1776 and died July 9, 1856. He was born and died in Turin, Italy. Amedeo Avodagro, conte di Quaregna e Ceretto, was born into a family of distinguished lawyers (Piedmont Family). Following in his family's footsteps, he graduated in ecclesiastical law (age 20) and began to practice law. However, Avogadro also was interested in the natural sciences and in 1800 he began private studies in physics and mathematics. In 1809, he started teaching the natural sciences in a *liceo* (high school) in Vericelli. It was in Vericelli that Avogadro wrote a *memoria* (concise note) in which he declared the hypothesis that is now known as Avogadro's law. Avogadro sent this *memoria* to De Lamétherie's *Journal de Physique, de Chemie et d'Histoire naturelle* and it was published.

In 1820, Avogadro became the first chair of mathematical physics at Turin University. In 1821, he published *Nouvelles considérations sur la théorie des proportions déterminées dans les combinaisons, et sur la détermination des masses des molécules des corps* and also *Mémoire sur la manière de ramener les composès organiques aux lois ordinaires des proportions déterminées*. In 1841, Avogadro completed and published his 4-volume work, *Fisica dei corpi ponderabili, ossia Trattato della costituzione materiale de' corpi*.

Not much is known about Avogadro's private life. He had six children and was reputed to be a religious man and also a discreet lady's man. Some historical accounts indicate that Avogadro sponsored and aided Sardinians planning a revolution on that island, stopped by the concession of Charles Albert's modern Constitution (*Statuto Albertino*). Because of his alleged political actions, Avogadro was removed as professor at Turin University (officially, the University was "very glad to allow this interesting scientist to take a rest from heavy teaching duties, in order to be able to give a better attention to his researches"). However, doubts remain as to the nature of Avogadro's association with the Sardinians. In any case, increasing acceptance of both revolutionary ideas and Avogadro's work led to his reinstatement at Turin University in 1833. Avogadro introduced the decimal system in Piedmont and served as a member of the Royal Superior Council on Public Instruction.

Avogadro's Law

Avogadro's law states that equal volumes of gases, at the same temperature and pressure, contain the same number of molecules. Avogadro's hypothesis wasn't generally accepted until after 1858 (after Avogadro's death), when the Italian chemist Stanislao Cannizzaro was able to explain why there were some organic chemical exceptions to Avogadro's hypothesis. One of the most important contributions of Avogadro's work was his resolution of the confusion surrounding atoms and molecules (although he didn't use the term 'atom'). Avogadro believed that particles could be composed of molecules and that molecules could be composed of still simpler units, atoms. The number of molecules in a mole (one gram molecular weight) was termed Avogadro's number (sometimes called Avogadro's constant) in honor of Avogadro's theories. Avogadro's number has been experimentally determined to be 6.022×10^{23} molecules per gram-mole.

Matrixes

Matrixes

With the following matrices a group of four equal in value integers represent a continuous continuation of that particular integer.

Matrixes

A matrix is a set of equations. We can solve for the unknown or for the various unknowns of the matrix. First look at some functions with matrices.

Addition & subtraction: Take an element of the first matrix and combine it with a corresponding element of the second matrix.

Multiplication: Here is a 2X 3 matrix times a 3X 2 matrix. The result will be a 2X2 matrix.

$$\begin{array}{ccc} & & 16 \quad -5 \\ -2 \quad 4 \quad -3 & \text{times} & 3 \quad 4 \\ 9 \quad 3 \quad 10 & & 12 \quad 9 \end{array}$$

THE First Half	The Second Half	The answer of the multiplication
$-2\cdot 16 + 4 \cdot 3 - 3 \cdot 12$	$-2(-5)+ 4 \cdot 4 - 3\cdot 9$	$=$ -56 -1
$9 \cdot 16 + 3 \cdot 3 + 10\cdot 12$	$9(-5) + 3 \cdot 4 + 10 \cdot 9$	273 57

To do matrix multiplication you are to take one row and multiply each element of that row by an element of a corresponding column. The results of the sum of each separate elements of a row times members of a column is to go in the corresponding column position of the row of where it started.

Next, we see a number of equations. We are to solve for the five unknowns which are A, B, C, D, and E.

1.) $0A + 3B - 3C - D + 2E = -1.90$
2.) $-3A - B - 3C + D + 0E = -5.10$
3.) $-2A + B + 0C + D + 2E = 0.40$
4.) $0A - 4B - 3C + 0D - 20E = 69.50$
5.) $0A + 0B + 2C - D + 3E = 31.60$

A.)   Next, the variables are removed and only the coefficients will remain. Also, the last column no longer will be presented after the equal sign/s.

1	+0	+3	-3	-1	+2	+1.9
2	-3	-1	-3	+1	+0	+ 5.1
3	-2	+1	0	+1	+2	- 0.4
4	+0	-4	-3	+0	-20	-69.5
5	+0	0	+2	-1	+3	-13.6

Here, again, we see five equations and there are five unknowns, A to E. Each unknown represents a dimension. Usually, in mathematics courses, there are only one, two, or three unknowns that represent one, two, or the three dimensions that we, humans, are familiar in our daily lives.

B.)   Because it is difficult to work out the
solutions when the number zero is in a position of
the diagonal, the First and Third Rows will be
exchanged, the Third Row will be divided by a
negative two, the Second Row will be divided by a
negative three, -3, the First Row will be divided
by a three, 3, the Fourth Row will be divided by a
negative four, -4, and the Fifth Row will be
divided by a two, 2.

3.)	1	-0.5	0	-.5	-1	0.2
2.)	1	0.3333	1	-0.3333	0	-1.7
1.)	0	1	-1	-0.3333	0.6666	0.63333
4.)	0	1	0.75	0	5	-17.375
5.)	0	0	1	-0.5	1.5	-6.8

C.)   Add a negative Row Three, -3, on to Row Two.

3.)	1	-0.5	0	-.5	-1	0.2
2.)	0	0.83333	1	0.16666	1	-1.9
1.)	0	1	-1	-0.3333	0.6666	0.63333
4.)	0	1	0.75	0	5	-17.375
5.)	0	0	1	-0.5	1.5	-6.8

D.)Divide Row Two, 2, by 0.83333.

3.)	1	-0.5	0	-0.5	-1	0.2
2.)	0	1	1.2	0.2	1.2	-2.28
1.)	0	1	-1	-0.3333	0.6666	0.63333
4.)	0	1	0.75	0	5	-17.375
5.)	0	0	1	-0.5	1.5	-6.8

E.)Add a negative Row Two, 2, on to row one, 1.

3.)	1	-0.5	0	-0.5	-1	0.2
2.)	0	1	1.2	0.2	1.2	-2.28
1.)	0	0	-2.2	-0.53333	-0.53333	2.913333
4.)	0	1	0.75	0	5	-17.375
5.)	0	0	1	-0.5	1.5	-6.8

F.)   Divide Row One, 1, by -2.2.

3.)	1	-0.5	0	-0.5	-1	0.2
2.)	0	1	1.2	0.2	1.2	-2.28
1.)	0	0	1	0.24242424	0.24242424	1.324242424
4.)	0	1	0.75	0	5	-17.375
5.)	0	0	1	-0.5	1.5	-6.8

G.)　　Add 0.45 of Row One, 1, on to Row Four, 4.

3.)	1	0.5	0	-0.5	-1	0.2
2.)	0	1	1.2	0.2	1.2	-2.28
1.)	0	0	1	0.24242424	0.24242424	1.324242424
4.)	0	0	0	0.090909090	3.90909090	-15.690909090
5.)	0	0	1	-0.5	1.5	-6.8

H.)　Divide Row Four by -0.09090909 and add a negative Row One, to Row Five.

3.)	1	0.5	0	-0.5	-1	0.2
2.)	0	1	1.2	0.2	1.2	-2.28
1.)	0	0	1	0.24242424	0.24242424	1.324242424
4.)	0	0	0	1	-43	172.6
5.)	0	0	0	-0.742424242	1.257575757	-5.475757575

I.)   Divide Row Five by 0.742424242.

3.)	1	0.5	0	-0.5	-1	0 . 2
2.)	0	1	1.2	0.2	1.2	-2.28
1.)	0	0	1	0.24242424	0.24242424	1.324242424
4.)	0	0	0	1	-43	172.6
5.)				-1	1.69387755	-7.375510207

J.)   Add Row Four on to Row Five.

3.)	1	0.5	0	-0.5	-1	0 . 2
2.)	0	1	1.2	0.2	1.2	-2.28
1.)	0	0	1	0.24242424	0.24242424	1.324242424
4.)	0	0	0	1	-43	172.6
5.)	0	0	0	0	-41.30612245	165.2344898

-41.30612245E = -165.2344898   So, E is equaled to 4.

D -43(4 or E) = -172.6

D   = -172.6 + 172

D   = -0.6

C + 0.24242424 + 0.24242424 -1.324242424 = 0

C + 0.24242424(-0.6 or D) + 0.24242424(4 or E) -1.324242424 = 0

C – 0.5 = 0

C = -0.5 + 0.5 = 0.5 This results when a 0.5 is added to both sides of the equation.

So, E =4, D= -0.6, and C = 0.5.

What do B and A equal?

B + 1.2 + 0.2 + 1.2 + 2.28 = 0

B + 1.2(0.5 or C) + 0.2(-0.6 or D) + 1.2(4 or E) -2.28 = 0   So

B   =   -3

E = 4, D = -0.6, C = 0.5, b = -3.

Last,

A – 0.5(-3 or B) + 0(0.5 or C) – 0.5(-0.6 or D) -1(4 or B) -0.2 = 0

So, A = 2.

E = 4, D = -0.6, C = 0.5, b = -3, A =2.

Matrix Division

Working with scalars and matrices is done by simply multiplying the elements and if the scalar is the inverse of an integer then the multiplication is similar to division. Yet, when doing matrix division many outcomes are possible. If an inverse is not something that can be found then the number of possibilities that are possible outcomes of the division is greater than one. So, we want to find an inverse when we are doing matrix division. Below is an example of matrix E time matrix F that result in Matrix G. With that known we can see that Matrix G divided by Matrix E should equal Matrix E. Also, the inverse of Matrix E, E^{-1}, times Matrix G will give us a result equal to Matrix E. So, multiplying G, which is the same as Matrix G, by E^{-1}, or the inverse of Matrix E, is similar to dividing G by E, where E is the same as Matrix E.

$$
\begin{array}{ccc}
E & F & G
\end{array}
$$

$$
\begin{vmatrix} 3 & -2 \\ 1 & 1 \end{vmatrix}
\begin{vmatrix} -1 & 1 \\ 4 & -8 \end{vmatrix}
=
\begin{vmatrix} -11 & 19 \\ 3 & -7 \end{vmatrix}
$$

To create the inverse of a matrix you should follow some key directions. The square matrix can be manipulated so that a multiple of each row can be added to another row in order to make almost all of the elements, summed with that addition, to be equaled to zero or to be equaled to one as a result. The numbers of the diagonal line, that starts at the top left corner of the square to the bottom right corner of the square, are to be manipulated to the point that the value of the elements of this line are all equal to the number one. As this is done, the neighboring identity matrix, a matrix with all zeros and a line of ones in the diagonal position, is treated in the same way as the matrix that is on the far right.

$$E^{-1} = \begin{vmatrix} 3 & -2 \\ 1 & 1 \end{vmatrix} \begin{vmatrix} 1 & 0 \\ 0 & 1 \end{vmatrix}$$

Divide the first row by three.

$$\begin{vmatrix} 1 & -2/3 \\ 1 & 1 \end{vmatrix} \begin{vmatrix} 1/3 & 0 \\ 0 & 1 \end{vmatrix}$$

Add the negative of row one to row two.

$$\begin{vmatrix} 1 & -2/3 \\ 0 & 5/3 \end{vmatrix} \begin{vmatrix} 1/3 & 0 \\ -1/3 & 1 \end{vmatrix}$$

Divide the second row by five thirds.

$$\begin{vmatrix} 1 & -2/3 \\ 0 & 1 \end{vmatrix} \begin{vmatrix} 1/3 & 0 \\ -1/5 & 3/5 \end{vmatrix}$$

Multiply the second row by two thirds and add the result to the first row.

$$\begin{vmatrix} 1 & 0 \\ 0 & 1 \end{vmatrix} \begin{vmatrix} 3/15 & 2/5 \\ -1/5 & 3/5 \end{vmatrix}$$

Now that the matrix on the far right is of an identity matrix format we know that the inversion is complete and the result on the far right is the inverse matrix, E^{-1}. Multiplying the inverse matrix by G will be the same as dividing Matrix G by Matrix E and the result should be Matrix F, because we have seen at the beginning of this section that E times F equaled G. Also, the inverse matrix times what it is the inverse of should bring out a result that is the same as an identity matrix.

$$\begin{vmatrix} 3/15 & 2/5 \\ -1/5 & 3/5 \end{vmatrix} \text{ times } \begin{vmatrix} 3 & -2 \\ 1 & 1 \end{vmatrix} = \begin{vmatrix} 1 & 0 \\ 0 & 1 \end{vmatrix}$$ This is an identity matrix.

So, then $E^{-1} G = F$

$$\underset{E^{-1}}{\begin{vmatrix} 3/15 & 2/5 \\ -1/5 & 3/5 \end{vmatrix}} \text{ times } \underset{G}{\begin{vmatrix} 11 & 19 \\ 3 & -7 \end{vmatrix}} = \underset{F}{\begin{vmatrix} -1 & 1 \\ 4 & -8 \end{vmatrix}}$$

Part Eleven Problems

11A.) $x^2/(2-x) + 5X = 6$, solve for X.

11B.) $3a - 2y = 6$

$-2a - 2y = -34$

11C.) $22X - 22Y + 0 = 12$

11D.) $X^2 - 9w + 15 = 6$

$X + 4w - 12 = 9$ Solve for the unknowns.

11E.) $34A + 2B - 4C - 4D = 22$

$21A + 4C = 33 - 2D$

$2A + 65 = 2C + 15C + 2^{-0.5}B$

$-8C - 8C = A + 18$

11F.) $88.89/0.00 =$ What?

11G.) $88.89^0 =$

11H.)   You have twenty-six U.S.A. coins that sum to a value two dollars and fifty-nine cents. How many of these coins are pennies, nickels, dimes, or quarters if you have less than six quarters.

11I.) $x(3) - X/2 = 4$, solve for X.

11J.) <u>Multiplication</u>

2	0	0	9	times	-2	22	9	-6
-2	-4	10	11		0	22	9	3
15	-2	8	22		-9	4	0	-2
0	-3	6	99		0	3	2	1

11K.) $X^2 - 9w + 15 = 6$

$X + w + 9 = 14$

11 L.) Solve the following matrix

$8A + 4B + 2C = -3$

$3A - 4B - C = 6$

$-3A - B + 2C = -3$

Part Twelve

The Train Man

Julian was born ten years after the start of World War Two. He dropped out of junior high school after being sent to prison for being caught with a burning marijuana cigarette in school. In prison he skipped school and served his time there with minor leisure that included much exercise. He learned wrestling, boxing, martial arts, basketball, and soccer. He was one of the jail's best chess players. Five years after being released he returned because he got into a big violent fight against his sister's verbally abusive husband. There fight was slow but intense. He used his pocket knife to cut the husband's pants seat, as the man he attacked wore the pants. After the fight, the evidence, a torn piece of pants, was used to enclose his life between bars, again. A prisoner release program arrived and he took advantage of it, where he had to work at collecting people's train tickets in the Cleveland, Ohio area. He loved the job, joined a church much later, and got married soon after joining the church.

Still, following rules was often broken by him. Fortunately, he broke some rules one day and forced a man to leave a chair that the man had taken from an old lady. When this man stood up about ten pounds of cocaine felled out of his jacket and a police officer was called by Julian to take the drug carrier off the train.

Later, on December 12, 1989, he witnessed and stopped a man who had entered the train with the wrong form of carfare. His carfare pass had expired. The man, an attorney, argued against Julian but he simply was handcuffed by Julian before the police was called by "the arresting train man." Before police arrived the attorney asked Julian could he make one telephone call. He was allowed one telephone call. On the train's telephone, he called his associates at "Howcome and We the Innocent Litigation" and told his partners that the train conductor handcuffed him because he had used illegal carfare. Their laughter was almost as loud as the train's momentum on the tracks until they paid the cost to have him released from jail and the court cost to have Julian removed from his job.

Because the attorney had, in the past, represented the man who Julian had caused to stand before the standing man dropped his cocaine, Julian's employer rehired him six months later and promoted him to a desk position where he monitored the train' system of protection cameras. Here, can Julian follow the rules without being over stressed and breaking them?

How did his wife react to the situation is another story?

Some Exponents

Why is X to the zero power, X^0, equaled to one and X to the negative one power, X^{-1}, equaled to the inverse of X, 1/X?

X^4/X^3 = 81/27 = 3^1 = X^1. So, when the fraction has a set of a known value or known values, such as 3, or an unknown or unknowns, such as X, where the exponents of the unknown values or known values differ or have the same value, we are to subtract one exponent value from the other exponent value. Therefore, $X^3/(X^1$ or X) is equaled to X^2, $X^3/X^2 = X^1$ which is the same as X, and $X^3/X^4 = X^{-1}$ or $1/X^1$ which is the same as 1/X because when the power is not shown we are to assume that the integer is one. Again, $X^3/X^4 = X^{-1}$. This is called the inverse function that can be written as 1/X or X^{-1}. Also, $X^2/X^3 = X^{-1}$, $X^1/X^2 = X^{-1}$, and $X^0/X^1 = X^{-1}$.

Next let's allow X to be equaled to two. Here, $2^2/2^1 = 2^1 = 2$, $2^2/2^2 = 1$, $2^2/2^3 = 1/2$ or 4/8 which is, after dividing the fraction's numerator and denominator by four, 1/2. So, 2^{-1}, which really is $(2/1)^{-1}$, equals 1/2 and X^{-1} equals 1/X. This is an inverse function, because the numerator and the denominator were exchanged.

This shows that X^0 equals one and X^{-1} is 1/X and it is both the inverse of X and the inverse of X/1. Please remember that all numbers have a denominator of one. Also, $X^{0.5}$ is the square root function, where $X^{1/2} = X^{0.5}$.

Questions

1.) Which fraction = the square root of X, A.) $X^5/X^{0.5}$, B.) $X^3/X^{3.5}$, or C.) $X^1/X^{0.5}$

2.) Write X to the one fifth power as an exponent.

Answers

1.) The answer to the first question is C.) Here, subtracting exponents gives us 1-0.5 = 0.5 and $X^{0.5}$ is the square root function. Choice A.) shows us $X^{4.5}$, and Choice B.) shows us $X^{-0.5}$ or $1/X^{0.5}$.

2.) The answer is $X^{0.2}$ because one power divided by five powers equals one fifth of a power, 0.2, or $X^{1/5}$. It also can be expressed as $1/X^{-0.2}$ because the negative exponent of the denominator equals a similar number in the numerator that is positive and has the same magnitude, absolute length or size. For example, the numbers -3 and +3, or 3, has the same magnitude which is three.

Formulas

1.)   Factorials

2.)   Economies of Scale and SEPTA

3.)   The Center of Gravity

4.)   The Sum of a Numerical Series

Good formulas are directions that lead us to correct answers or creations.

November 10, 2008

Mandela leads tributes to 'Mama Afrika', songstress Miriam Makeba

Fred Bridgland in Johannesburg

'Mama Afrika' dies at Italian anti-Mafia concert

Nelson Mandela was among thousands of South Africans to pay tribute today to the singer and activist, Miriam Makeba, who died suddenly after taking part in a concert against the Italian Mafia.

Her death provoked shock and widespread mourning in a country enchanted by the sweetness and shining sound of her singing.

Mandela, now in his 91st year and who rarely makes public statements any more, led the tributes to Makeba. "She was South Africa's first lady of song and so richly deserved the title of Mama Afrika," he said.

Related Links

- Miriam Makeba dies at anti-Mafia concert

- ANC 'terrified' of music claims apartheid hero

Multimedia

"The sudden passing of our beloved Miriam has saddened us ... For many decades, starting in the years before we went to prison, MaMiriam featured prominently in our lives and we enjoyed her moving performances. When she went into exile she continued to make us proud as she used her worldwide fame to focus attention on the abomination of apartheid. Her music inspired a powerful sense of hope in all of us. She was a mother to our struggle and to the young nation of ours.

"It was fitting that her last moments were spent on a stage, enriching the hearts and lives of others - and again in support of a good cause."

Relatives and friends who first encouraged Makeba to sing compared her voice to that of a nightingale. Her distinctive style, which bewitched the world in the 1960s and 1970s, combined traditional African melodies, jazz and folk with the unique and dynamic rhythms of South Africa's black townships.

While she toured with Harry Belafonte and sang with Marilyn Monroe at John F. Kennedy's birthday party at Madison Square Garden in 1962, her music was banned in South Africa by apartheid governments. When she first travelled to New York in 1960 to perform with Belafonte, the Pretoria government refused to allow her to return home.

She lived in exile for the next 31 years. Mandela asked her to come home after his release from life imprisonment in February 1990 and when she arrived in Johannesburg she said: "I never understood why I couldn't come home. I never committed any crime."

Makeba collapsed shortly after a performance in the southern Italian town of Castel Volturno yesterday evening and died in hospital early today. She was paying homage to six Africans killed by the Camorra mafia two months ago and to the Italian journalist Roberto Saviano who exposed the murders and was himself threatened with death.

South Africa's foreign minister Nkosazana Dlamini-Zuma said in a statement: "One of the greatest songstresses of our time has ceased to sing. Throughout her life, Mama Makeba communicated a positive message to the world about the struggle of the people of South Africa and the certainty of victory over the dark forces of apartheid and colonialism through the art of song."

Makeba's body is being flown back to South Africa for a funeral and burial in Johannesburg.

Makeba's career soared in America and Europe until 1968 when she married the black activist Stokeley Carmichael, "Honorary Prime Minister" of the Black Panther Party. She was in immediate trouble with the FBI and all her American concerts and recording contracts were cancelled.

The couple moved to the West African state of Guinea-Conakry, ruled by a dictator, Ahmed Sékou Touré, who imprisoned political opponents in camps, such as the notorious Camp Boiro National Guard Barracks, and drove tens of thousands of dissidents into exile. Carmichael took the name Kwame Touré. Makeba was given the resources to develop a distinctive West African style of music while representing her new country at the United Nations.

During this period in Sékou Touré's state, Makeba virtually disappeared from international view. But after her divorce from Carmichael and the death of her only child, her daughter Bongi, in 1985 she settled in Brussels and began performing to international audiences again. She remained popular, but the sheer sweetness of her young voice was gone.

Makeba's publicist Mark Lechat said the singer had suffered from severe arthritis and had been unwell for some time, appearing at concerts with the aid of a stick. She was married four times. One of her husbands was the trumpeter Hugh Masekela.

1.) <u>FACTORIALS</u>   <u>AND MORE!</u>

Here and in following lessons multiplication will be indicated by a dot between two numbers or a set of brackets between numbers. Note: X^n is X to the nth power. Examples $5^4 = 5 \cdot 5 \cdot 5 \cdot 5 = 625$ and $11^3 = 1,331$.

Here are just two Number Facts and four questions.

A.)<u>Factorials</u>

Factorials are a unique type of operation. They can only be done on positive integers, ex. 0, 1, 2, 3, 4, 5, etc. An example is seven factorial,

$$7! = 7 \cdot 6 \cdot 5 \cdot 4 \cdot 3 \cdot 2 \cdot 1 = 5,040.$$

Here, the integer number, in this case seven, is multiplied by the integer that is one integer less than that starting and largest integer number. This is continued until the set of integers includes the number one. Of course one can purchase a calculator and use the factorial button to find the answer for seven, or any other, factorial. Yet, now you know how to derive the answers to factorial questions.

Seven factorial divided by six factorial is

$7!/6! = \dfrac{7 \cdot 6 \cdot 5 \cdot 4 \cdot 3 \cdot 2 \cdot 1}{6 \cdot 5 \cdot 4 \cdot 3 \cdot 2 \cdot 1}$ After canceling numbers that are both in the numerator and the denominator is 7.

$6!/5! = \dfrac{6 \cdot 5 \cdot 4 \cdot 3 \cdot 2 \cdot 1}{5 \cdot 4 \cdot 3 \cdot 2 \cdot 1}$ After canceling numbers that are both in the numerator and the denominator is 6.

$5!/4! = \dfrac{5 \cdot 4 \cdot 3 \cdot 2 \cdot 1}{4 \cdot 3 \cdot 2 \cdot 1}$ After canceling numbers that are both in the numerator and the denominator is 5.

$4!/3! = \dfrac{4 \cdot 3 \cdot 2 \cdot 1}{3 \cdot 2 \cdot 1}$ After canceling numbers that are both in the numerator and the denominator is 4.

$3!/2! = \dfrac{3 \cdot 2 \cdot 1}{2 \cdot 1}$ After canceling numbers that are both in the numerator and the denominator is 3.

$2!/1! = \dfrac{2 \cdot 1}{\cdot 1}$ After canceling numbers that are both in the numerator and the denominator is 2.

Following this pattern of answers, from seven factorial to two factorial, the next answer should be one factorial. So, as what has been done up to now, the numerators and the denominators are reduced by one and the result should be one. Thus, zero factorial equals one.

$1!/0! = \dfrac{1}{1}$ After canceling numbers that are both in the numerator and the denominator is 1.

Note: In respect to a man who arrived to the U.S.A. from Zambia, who is a mathematics teacher, this proof is not agreed to by everyone. Yet, although he told me that I am wrong and wrote a good book about the number zero, he never explained why is zero factorial equal to the number one. We were disagreeing "tough" until Fela Anikulapo Kuti's music, which we two non-Nigerians liked very much, came on and we like that type of sound. Dude got terminated from his job as a teacher in the Philadelphia, Pennsylvania School Board because when he taught students about the low chances of winning any game of chance some told their parents that the teacher was trying to get them to gamble. From there he lost his job and found more, too much, alcohol to entertain himself.

Then he met me in a shelter and I was not about to give up on my reason for why zero factorial equals one.

B.) Possible Arrangements

1.) The number of possible arrangements of one item is what?

The answer is that there is only one possible arrangement.   For example, if you have one letter named A there is only one arrangement for this set. We can call this 1! and this equals one.

2.) The number of possible arrangements of two items is what?

The answer is that there are only two possible arrangements. For example, if   you have two letters named AB there are only two arrangements for this set. We can call this 2! and this equals two, AB and BA.

3.) The number of possible arrangements of three items is what?

The answer is that there are only six possible arrangements. For example, if you have three letters named ABC there are only six arrangements for this set. We can call this 3! and this equals six.   ABC, ACB, BAC, BCA, CAB, & CBA

4.) The number of possible arrangement/s for zero is what?

Since there is only one item then there is only one arrangement. Thus, again, zero factorial equals one.

5.) Last, for today and so that this does not drive you crazy,

the number of possible arrangements of four items is what?

The answer is that there are only twenty-four possible arrangements.

For example, if you have four letters named ABCD there are only twenty-four

arrangements for this set. We can call this 4! and this equals twenty-four.

BACD, ABDC, ACBD, ACDB, ADBC, ADCB   BACD, BADC, BCAD, BCDA, BDAC,BDCA

CABD, CADB, CBAD, CBDA, CDAB, CDBA   DABC, DACB, DBAC, DBCA, DCAB, DCBA

38.) How many ways can we arrange the integers of 1382491?

39.) What is the result of $5^{3!}$?

40.) -66.08 + 43.23 – 4.09 + 1,008.19 =

41.) 34-54 = 66 + X. What is the value of X?

42.) With four decimal places in your answer, if a circle has a diameter value of 8 feet what is the length of X when the length of Y is 3?

43.) -66.087 –140 + 3.37 =

44.)   34-54 = 66 + X. What is the value of X?

45.) With four decimal places in your answer, if a circle has a diameter value of 18.1 meters what is the length of Y when the length of X is 0.5?

46.) ln 64/ln16 =

55555555555555555555555555555555555555555555555555555555555

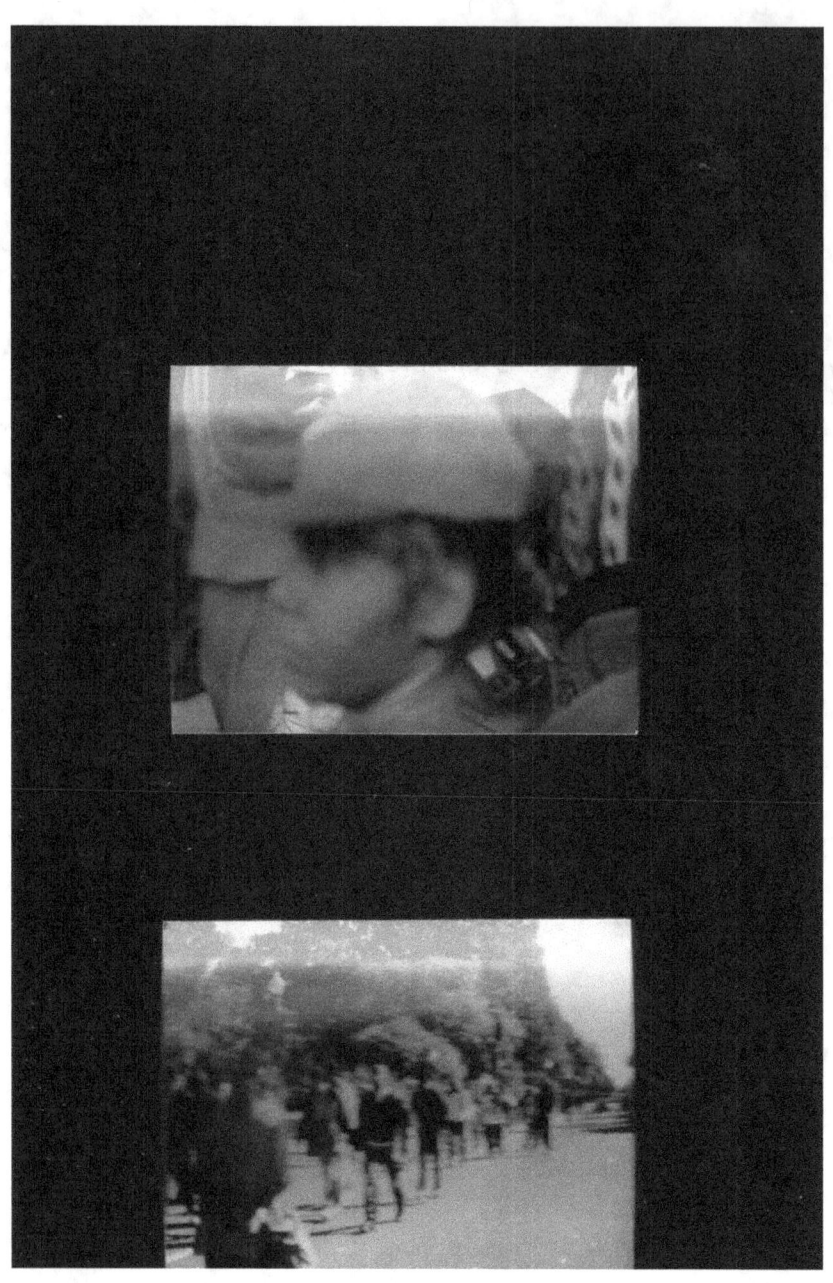

Part Twelve Problems

12A.) $36^{2.5}$ = What?

12B.) What is the value of (12 -2) Factorial?

12C.) 3 factorial divided by 12 factorial =

12D.) 6! =

12E.) 4(10!/8!)^2 – 19.

12F.) (2-X) + 5X = 15 Solve for X.

12G.) How many different ways can seven different types of fruit be placed in a line of fruits?

12H.) Find twelve numbers that differ by three, 3, and sum to forty-five, 45. See the next section for instructions.

12I.)   Find four numbers that differ by sixteen, 16, where the first number is twenty-four, 24. These numbers will sum   up to what value.

12J.)   Solve the following matrix.   15K – 24L + 3.4M + 22 = -91.00

-3K – 22L -8M   + 34.4 = 0.50

2K -50L +3M   + 10   = 1.00

Part Thirteen

2.) Economies of Scale

How many Philadelphia SEPTA transpasses can a passenger lose and still find savings, or break even, after purchasing Philadelphia monthly SEPTA transpasses instead of Philadelphia weekly SEPTA transpasses?

We need to design a system of facts and use these facts to find the answer. First, the price of a weekly transpass, today, is twenty dollars and seventy-five cents, $20.75. Second, the price of a transpass, today, is $83.00. Also, we need some time frame. Let's limit the time to be considered to be one year. Last, let us assume that if one can lose a monthly transpass one can also lose a weekly transpass. Also, let us, for now, assume that during times of when transpasses are lost, the day is both the first day that weekly transpasses become activated and the first day of the month, which is when monthly transpasses become activated. Here, X is the number of times one can lose a transpass, fifty-two is the number of weeks in a year for one to purchase a weekly transpass, and twelve is the number of months in a year for one to purchase a monthly transpass.

$$(52 + X)(\$22.5) - (12 + X)(\$83) = 0$$
$$(1{,}170 + 22.5X) - (996 + 83X) = 0$$
$$1{,}170 - 996 = 83X - 22.5X$$
$$174 = 60.5\,X$$
$$174/60.5 = X = 2.876033058$$

The above numbers make the following system.

Let X be the unknown number of times that one can lose a monthly transpass and break even by purchasing a monthly transpass instead of a weekly transpass.

Let W be the yearly cost of weekly transpasses.
Let M be the yearly cost of monthly transpasses.
Let w be the cost of a weekly transpass.
Let m be the cost of a monthly transpass.

$$(52 + X)(w) - (12 + X)(m) = 0$$
$$W + wX - M - mX = 0$$
$$W - M = mX - wX$$
$$W - M = X(m - w)$$
$$\{(W - M)/(m - w)\} = X$$

Another way of looking at X is to see that it is the yearly expenditure difference between the price of a year's worth of monthly transpasses and a year's worth of weekly transpasses divided by the difference in the price of a monthly transpass and a weekly transpass.

What happens after we leave the "abstract world" and look at the "real world?"

$174/$60.5   = X =   2.876033058 which equals more than two or two and one one-half times a year that one can lose a monthly transpass and still break even by purchasing monthly transpasses instead of purchasing weekly transpasses. Are the savings that one gets by purchasing a monthly transpass, instead of purchasing weakly transpasses, more or less than they were before SEPTA, Southeastern Pennsylvania Transportation Authority, increased its prices?

This shows us that one can lose more than two monthly transpasses a year and still break even, or not lose more than all of the funds saved by purchasing monthly transpasses instead of purchasing weekly transpasses. This exercise proves that, although many have said that I don't want to buy a monthly transpass because I might lose it, even if one loses two monthly transpasses per a year he or she will still benefit by buying monthly transpasses instead of weekly transpasses.

<u>Notes</u>

3.) <u>The Center of Gravity</u>

<u>Some unusual number line exercises</u> Here are two number line exercises.

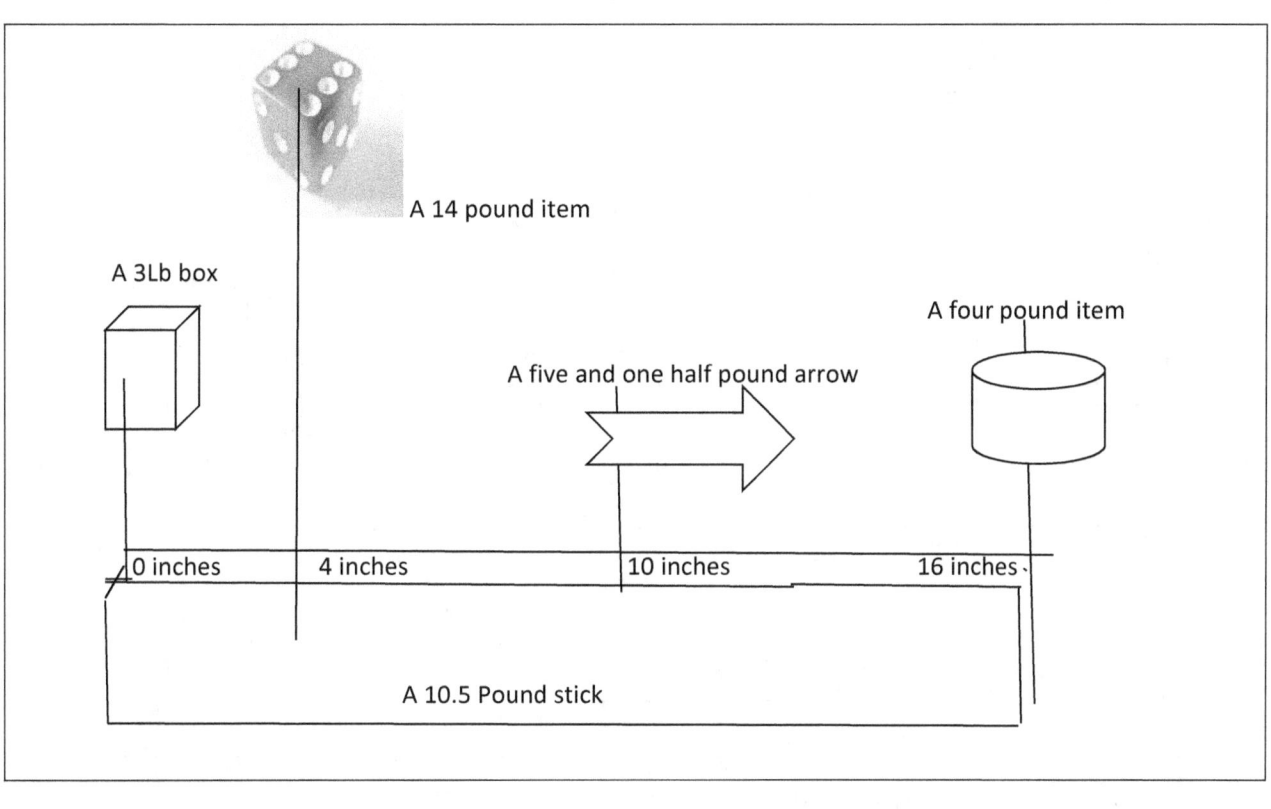

The fifth figure is the six pound box on the left.

A 6 pounds item

    A stick can be balanced if it rest at a spot on its center of gravity. In this number line exercise we will balance some items on imaginary sticks or lines.

    In the first example, the weights of four items have been placed near the above figures. Where should the fifth figure, that has a six pound weight, be placed on the ten and one-half pound stick that holds all of the four figures so that the items will be balanced around the origin?

First, find the center of the stick, or line, and call that center zero. Left of it is the number line's negative area and right of this origin is the number lines' positive area. The origin occurs eight inches from the left or right hand side of the stick. Next, multiply the distance that each item is from the origin, zero, by the weight of the item. Sum these values and divide the sum by the total weight of the stick and the various items.

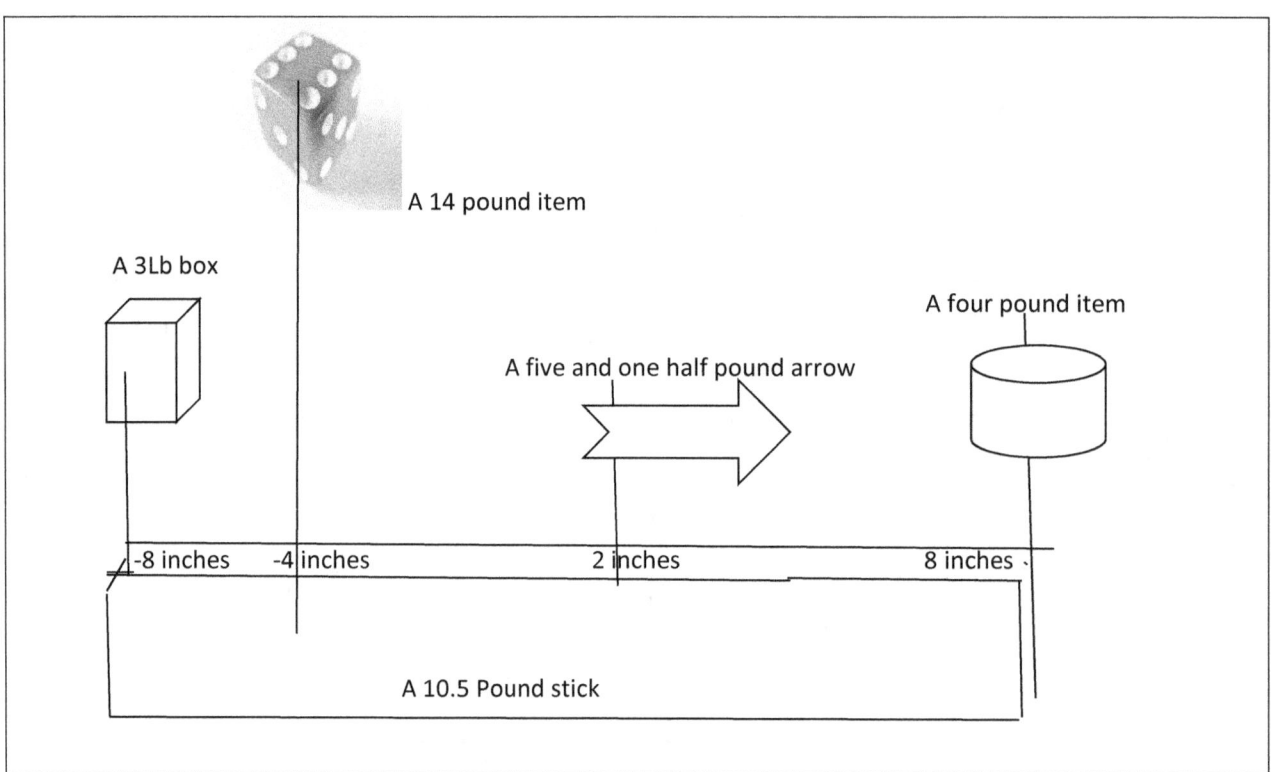

A 14 pound item

A 3Lb box

A four pound item

A five and one half pound arrow

-8 inches   -4 inches   2 inches   8 inches

A 10.5 Pound stick

The fifth figure is the six pound box on the left.

A 6 pounds item

$$\frac{-24 - 56 + 11 + 32}{3 + 14 + 5.5 + 4 + 10.5} = \frac{-37}{37} = \boxed{\text{-1.00 from the origin is the center of gravity of this system.}}$$

Should we place the fifth item on 6.1666 the weight of the items will be balanced. To find the place to place this figure all we had to do was to find out what had to be done to make the numerator be zero. So with the opposite sign of the value of -37, this number was divided by the weight of this fifth item and the result is 6.1666.

$$\frac{-24-56 +11+ 32 + 37}{3+14 + 5.5+ 4+10.5 + 6} = \frac{0}{43} = 0.00 \text{ from the origin.}$$

4.) __The Sum of a Series of Numbers__ that differ by a particular Number

The first number of these series is represented by the letter capital F, the difference between numbers of the series is represented by the symbol delta or Δ for change, and the octothorpe symbol or # represents the number of items in the series of numbers.

$$(\#)\left[F + \Delta(\# - 1)/2\right]$$

The following two series are examples of how to use this formula.

19, 22, 25, 28, 31, 34, 37   and   -37, -22, -7, 8, 23, 38, 53, 68
In the first example the first number is nineteen, the difference between numbers, Δ, is three, and the number of events is seven. So the sum of these numbers can be found by the above formula. In the second example the first number is negative thirty-seven, the difference between numbers is fifteen, and the number of events is eight.

$$(7)[19 + 3(7 - 1)/2] = 196 \quad \text{and} \quad (8)[-37 + 15(8-1)/2] = 124$$

To find these answers we should remember the poem
    Please Excuse My Dear Aunt Sally.

This really stands for parenthesis, exponents, multiplication, division, addition, and subtraction. These six items should be done in the order of the words of the above poem.

Problem 12H.) is <u>Find twelve numbers that differ by three, 3, and sum to forty-five, 45.</u>

Here, we can use the above formula, (#)[F + Δ(# - 1)/2].

(12)(F + 3[(12-1)/2] = 45

(12)(F + 16.5] = 45

(F + 16.5) = 3.75

F = - 12.75

1	-	12.75
2	-	9.75
3	-	6.75
4	-	3.75
5	-	0.75
6		2.25
7		5.25
8		8.25
9		11.25
10		14.25
11		17.25
12		20.25

Part Thirteen Problems

(Do all multiplication, division, addition, and subtraction without use of a calculator.)

13A.) Where do the functions 3X -2 and -4X + 8 meet?

13B.) At X = 0 a triangle has one point at Y = 5, at X = 25 Y is at 25, and
at X = 5 Y is at -0.2. What is the area of this triangle?

13C.) Triangle Cc is located at (-16,5), (22,-1), and (-4, 0). What is the area
of this triangle?

13D.) Six quarters are to be placed four inches from an edge of a yard
stick, that does not have any mass, and there are five more quarters
available. Where should they be placed to make this yard stick
balance on a string that is perpendicular to this stick and that is
three feet above the ground?

13E.) A circle has a diameter of 17.585 centimeters. This circle's
circumference is what?

13F.) In question 13D where is the center of gravity if the string has a
weight of 0.16 pounds?

13G.) $[(15^{-2})(15^3)/23 - 9]$ = ?

13H.) Hanging from a string a three pound box is twenty feet from a sixteen
pound case that is hanging from the same string. The string is 84 feet
long and on a balance. Where is the center of gravity in this system?

13I.) Factor: $18X^2 - 3X^4 + 21X$

13J.) Multiply the following items; $(27X^2 -3X +4)(6X -8)$.

13K.) Find five numbers that differ by three and one-half and sum to forty-
five.

D.C. People

Part Fourteen

Circles and Spheres

Find the distance between two circles. Here, Circle A has a radius of four meters and Circle B has a radius of seven meters. The center of Circle A is (-6, -15) and the center of circle B is at (5, 30).

The answer is found by using the Pythagorean Theorem. Here, we find the distance between the two centers and subtract from that the two different radius values.

$$([\,5 - -6]^2 + [\,30 - -15]^2)^{0.5} - (4+7) = 35.3249$$

———————————

If one figure is Sphere B with a center at (5, 30, 13) and the other is Circle A with a center at (-6, -15, 8), what is the distance between the sphere and the circle if the radius of the sphere is 7 and the radius of the circle is 4? $([\,5 - -6]^2 + [\,30 - -15]^{2+} + [13-8]^2)^{0.5} - (4+7) = 35.5940$ $(X^2 + Y^2)^{0.5}$ is of the Pythagorean Theory. What happens to this relationship when the Z axis is introduced? Here, the formula becomes $(X^2 + Y^2 + Z^2)^{0.5}$.

Why?

In the XY plane Pythagorean Theory we found the length of line H, or the hypotenuse, by solving $(X^2 + Y^2)^{0.5}$. After getting the answer we can use it to find the distance from the origin, not to the opposite rectangular corner but, to the opposite three dimensional corner. This can be done by noting that we can consider the hypotenuse, line h, to be one of two important lengths, just as X was one of two important lengths and Y was the other one of two important lengths when we were trying to find the length of the hypotenuse, line h. This time h will be one of the two important lines and Z will be the other of the two important lines for us to find the length of the three dimensional line H. So we will use

$(h^2 + Z^2)^{0.5}$ to get the distance from the origin, or center, of one of the two figures to the origin, or center, of the other of the two figures. Note that the formula
$(h^2 + Z^2)^{0.5}$ Is really $[(X^2 + Y^2)^{0.5} + Z^2]^{0.5}$ and after some calculating it is $[X^2+Y^2+Z^2]^{0.5}$.
Again, h= $[(X^2 + Y^2)^{0.5}$, so, $h^2 = [(X^2 + Y^2)^{0.5}]^2$ or $= (X^2 + Y^2)$. So, $(h^2 + Z^2)^{0.5} = (X^2 + Y^2 + Z^2)^{0.5}$.

If Circle A turned into being a sphere where Sphere A is a figure with a center at (-6, -15, 8) and Sphere B is a figure with a center at (5, 3, 13), what is the distance between the spheres if the radius of the Sphere B is seven and the radius of the Sphere A is four ?   ([5 - - 6]² + [3 - - 15]² + [13 -8]²)$^{0.5}$ – (7 + 4) = 10.6795, when four decimal places are used right of the decimal point.

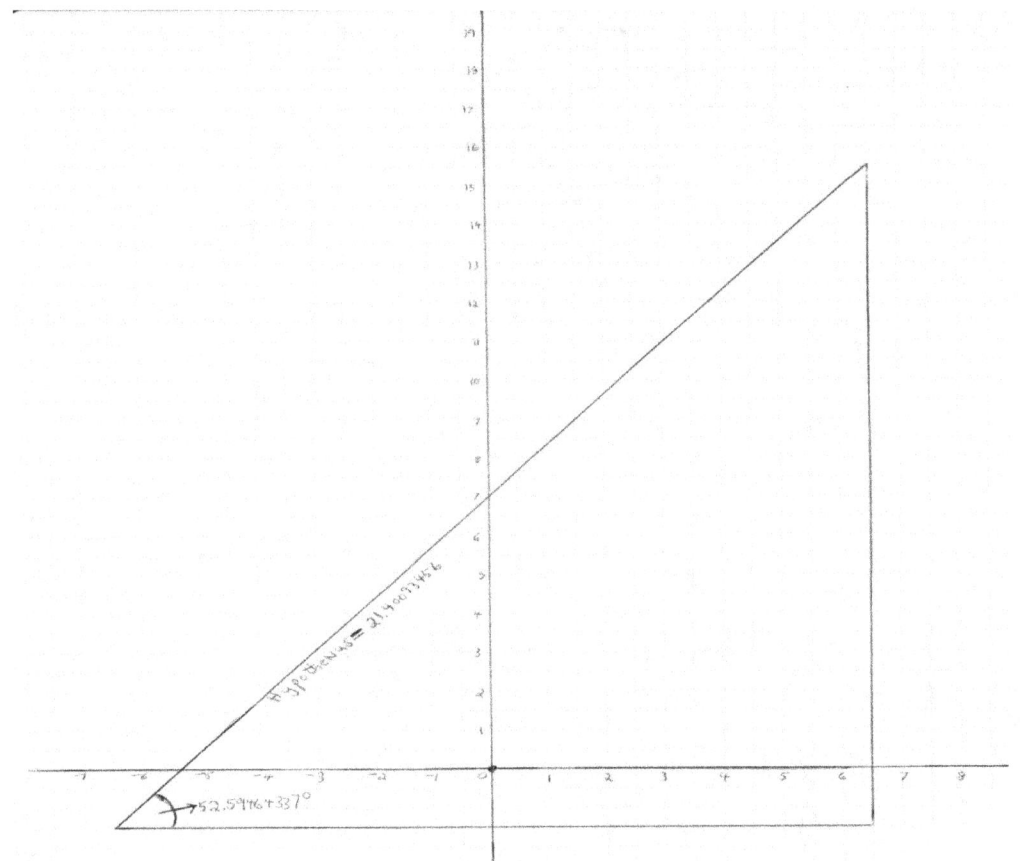

The angle on the far right and bottom hand side of this triangle is an angle that measures 90 degrees. So, the triangle is called a right triangle. Since all triangles have 180 degrees, the unlisted angle must have a measurement of 37.0535663 degrees.

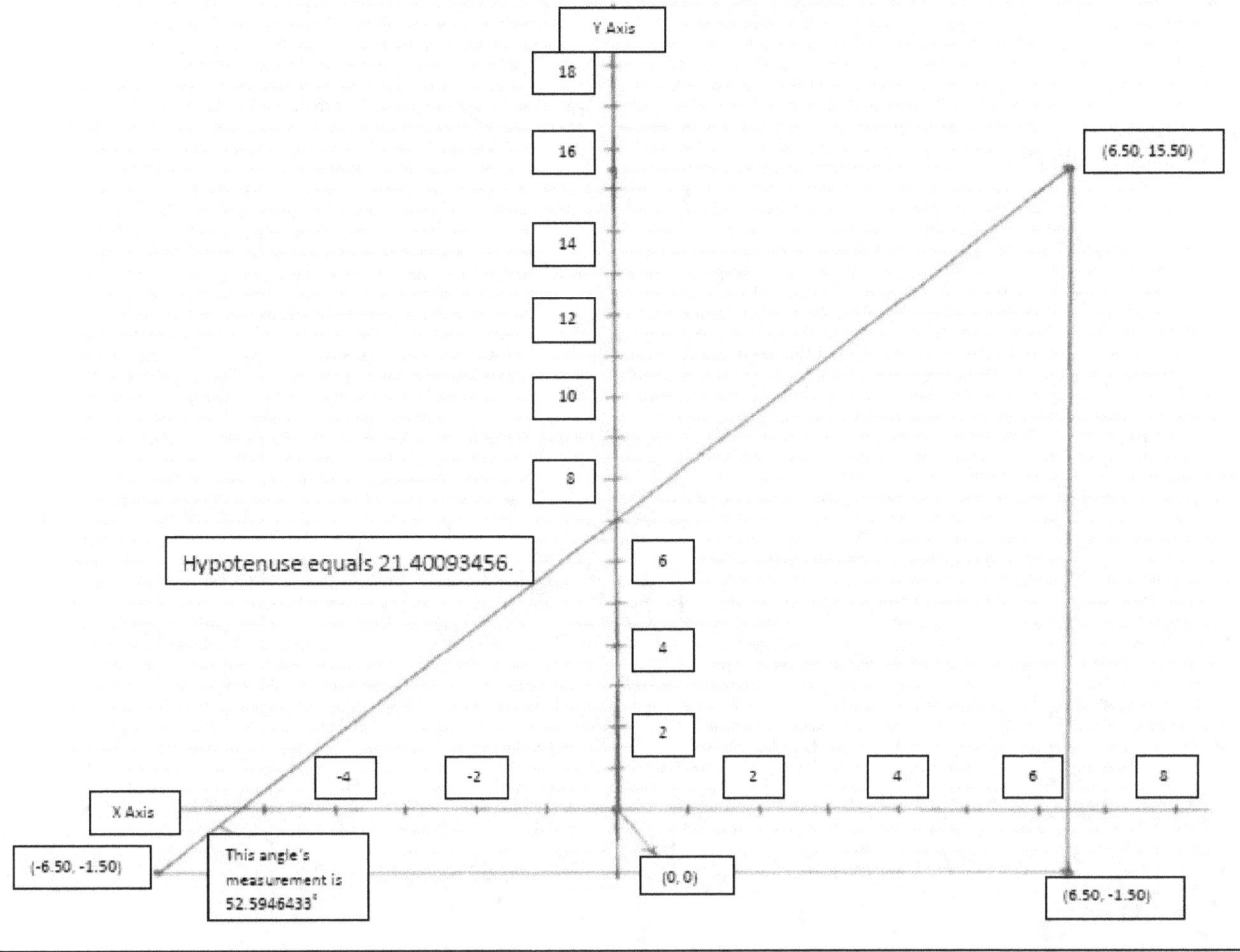

The above triangle has a hypotenuse of 21.40093456 because $(13^2 + 17^2)^{0.5} = 21.40093456$. Also, the sin equals the opposite side of the angle divided by the hypotenuse of the triangle so the angle equals the arc^{-1} sin of 17/ 21.40093456 and it is 52.5946433°, the cosine equals the adjacent side of the angle divided by the hypotenuse of the triangle so the angle equals the arc^{-1} cos(13/21.40093456) = 52.5946433°, and the tangent equals sin of the angle/cos of the angle or the (opposite/hypotenuse)/(adjacent/hypotenuse) which is the (opposite side)/(the adjacent side). Here, you should first type in 17/13 then press in the tan^{-1} key to get the angle of the angle, 52.5946433°.

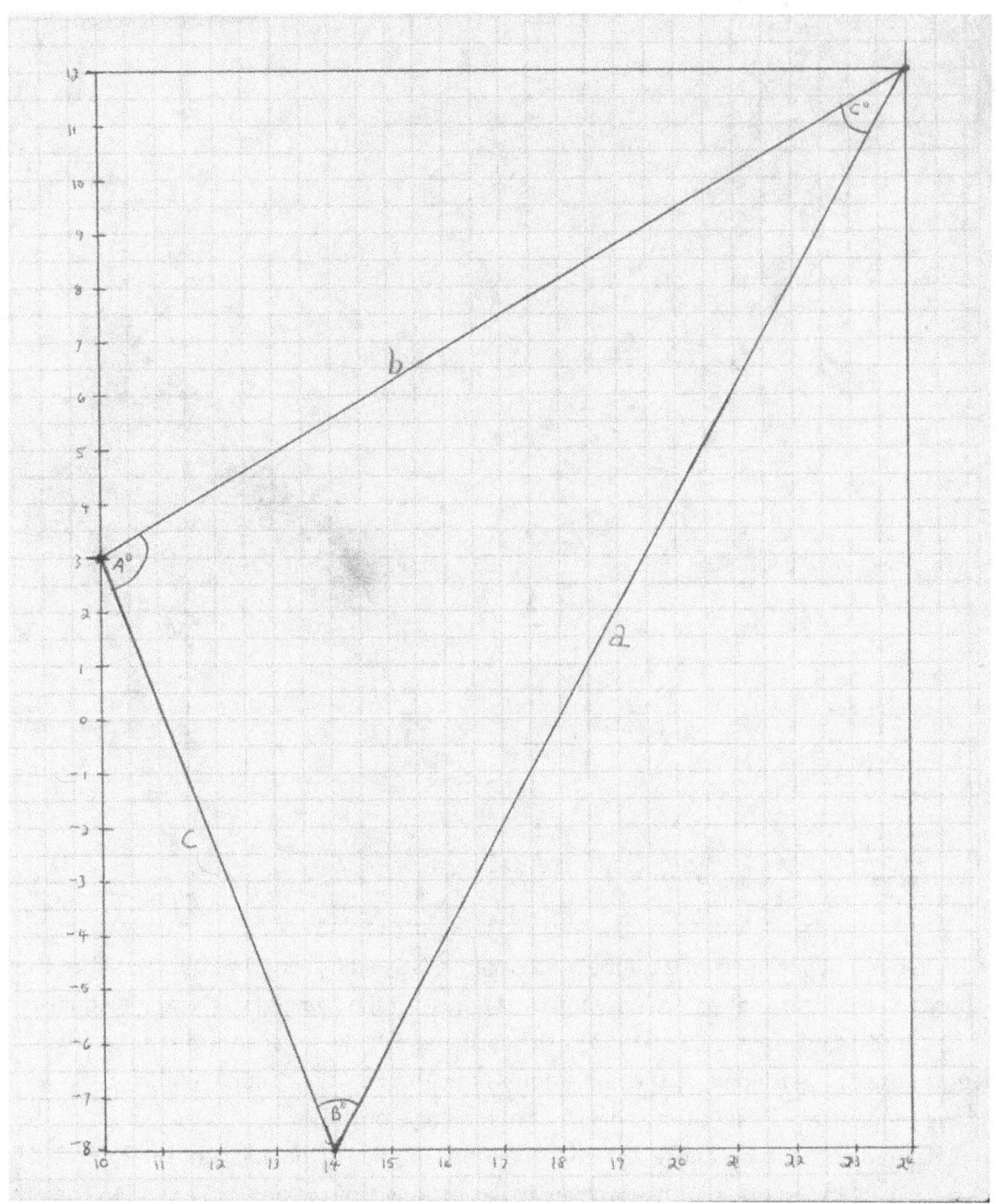

You Will See Me, Later

The Avenue

Notes

_____

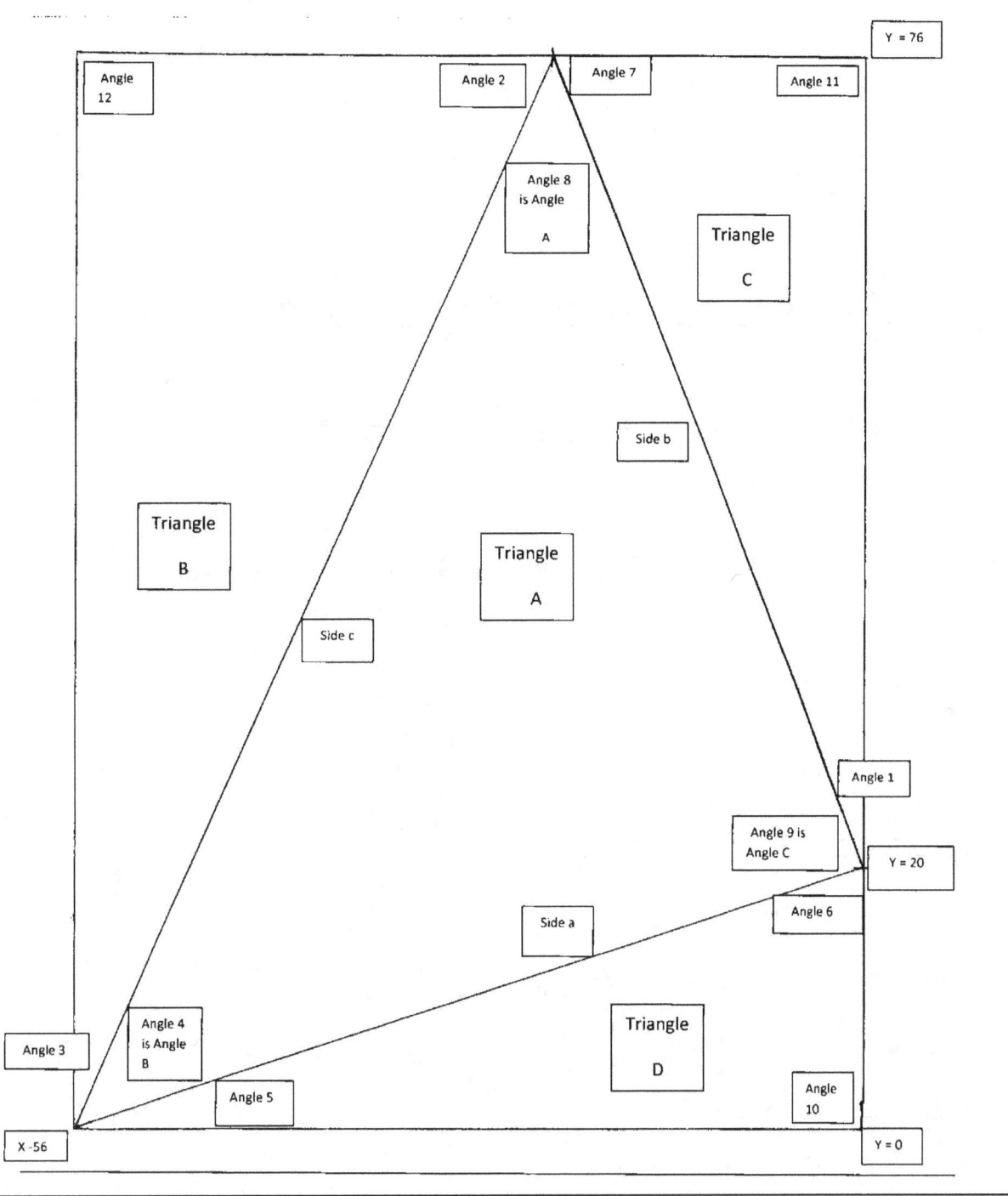

Find the area of triangle A.
   Show the results of the
        1.) Law of Sines
        2.) Law of Cosines and
        3.) Herons' Formula

(Some) Trigonometric Identities

Sine   =   Opposite/Hypotenuse

Cosine   =   Adjacent/Hypotenuse

Tangent   =   Sine/Cosine

Cotangent   =   Cosine/Sine

Secant   =   1/Cosine

Cosecant   =   1/Sine

Heron's Formula.

Here SS represents "Sum of the Sides."

$$((SS)/2[(SS)/2 - a] \ [(SS)/2 - b] \ [(SS)/2 - c])^{0.5}$$

Use Heron's formula to find the area of this triangle and show that it has the same value as what we will find by using a method that uses trigonometry and geometry to find the value of the area of this triangle.

01.) Side "a" equals the square root of the sum of twenty squared plus fifty-six squared. Here, this is the Pythagoras theorem that states that the hypothesis of a triangle equals the sum of the square of the two other sides of a triangle.

Side A = 59.46427499 or $(20^2+56^2)^{0.5}$

Side B = 60.92618485 or $(24^2+56^2)^{0.5}$

Side C = 82.46211251 or $(32^2+76^2)^{0.5}$

Adding up the areas of triangles B, C, and D will result in a value that will be temporarily called "I." The area of the rectangle is (76)(56), or the length of the two sides. From that subtract the value of the sum of the triangles B, C, and D, and the result has to equal what is left, the value of triangle A.

```
Rectangle's area    =  (76)(56)    =    4,256
          MINUS
Area of Triangle B = (76)(32)/2    =    1,216
Area of Triangle C = (56)(24)/2    =      672
Area of Triangle D = (56)(20)/2    =      560
                                        ─────
                                        1,808
```

Let us do a few more things with this problem.

To prove that the value of the area of triangle A is 1808 we will observe the value by using

a.) The Law of Sines and Cosines.
b.) Heron's Formula

Note that the term arc sine is also known as $\sin^{-1}$ and it is the angle whose sine has a particular value.

Follow along and you will see good examples of the arc-sine function.

Proof from using
The Law of Sines and Cosines

Here, is the Law of Sines.
a/Sin A = b/Sin B = c/Sin C
where a, b, and c are sides of a triangle and A, B, and C are angles of that same triangle.

The law of Sines can be used to find angles of a general triangle. If two of the triangle's sides and the angle that is enclosed by them are known, then the law of Sines can be used with the law of cosine to find the length of the third side of the triangle and the other two angles of that triangle.

Here, is the Law of Cosines.
a^2 $= b^2 + c^2$ −2bc Cos A
b^2 $= a^2 + c^2$ −2ac Cos B
c^2 $= a^2 + b^2$ −2ab Cos C

01.) First we will observe the values of the angles
     of the rectangle.

02.) Second we will check to see whether the values
     of the angles of Triangle A add up to a value
     of 180 degrees.

     Angles
     01
     02
     03
     04
     05
     06
     07
     08
     09
     10   This right angle is 90°
     11   This right angle is 90°
     12   This right angle is 90°

Let's observe the value of the angles of triangle B.
To do so, let's observe the value of angle 2. Note
that each horizontal and each vertical line is
separated by a value of two between it and the next
parallel line. The length of line B must be the
square root of the sum of $X^2 + Y^2$, or $(24^2 + 56^2)^{0.5} =$
60.92618485.

The sine of any angle, including Angle 2, is

$$\frac{\text{The Length of the Opposite Side}}{\text{The length of the Hypothesis}}$$

Angle 2 = the arc sine of 76/82.46211251 =
              67.16634583 degrees or 67.16634583°.
Angle 3 = the arc sine of 32/82.46211251 =
              22.83365418 degrees or 22.83365418°.

Right Angles and Angles of Triangle B

    01
    02 67.16634583°
    03 22.83365418°
    04
    05
    06
    07
    08
    09
    10 This right angle is 90°
    11 This right angle is 90°
    12 This right angle is 90°

Let's observe the value of the angles of triangle c.
To do so, let's observe the value of Angle 1.
To find the hypotenuse we summed the square of
sides X and Y before getting the square route of
that sum.

$$(24^2 + 56^2)^{0.5} = 60.92618485,$$

which is side b.

The sine of any angle, including Angle 1, is

The Length of the Opposite Side

The length of the Hypothesis

This is   24/60.92618485 or 23.32329182

        IF YOU THINK I'M A LIAR THEN CHECK THE
NUMBERS ABOVE WHERE THE LENGTH OF SIDE B IS
LABELLED AS 60.92618485 AND IN THE PICTURE OF THE
TRIANGLE WHERE SIDE A IS LABELLED AS BEING 24.
    O.K….
Moreover, if you want a SNACK, you have to do the
math.

And to find the angle that has this value, get the
arc sine of this value. This will give you the
value of the angle. The arc sine of 0.3939 is
the value of angle 1.

Angle 1 =   23.19738752°

We know the value of angle 11. It, again, is a
right angle with ninety, 90, degrees. Next to find
all of the values of triangle C, we need to observe
the value of angle 7.

Angle 7 = Opposite/hypothesis or
    56/60.92618485 = 0.91914503.
    The arc sine of this is 66.80261248°.
    Yet, for convenience, we will it can be called
    as being 66.8° or

Angle 7 =   66.80261248 degrees.

The sum of the three angles of this triangle C must add up to equal 180
degrees and they do that.

Right Angles and Angles of Triangle C

```
01   23.19738752°
02
03
04
05
06
07   66.261248°
08
09
10   This right angle is 90°
11   This right angle is 90°
12   This right angle is 90°
```

Right Angles and Angles of Triangle D.

```
01
02
03
04
05   19.65382406°
06   70.34617594°
07
08
09
10   This right angle is 90°
11   This right angle is 90°
12   This right angle is 90°
```

° The circle represents the word degree/s.

Angle 5 is found by getting the arc sine of
20/59.46427499, which is 19.65382406°.

Angle 6 is found by subtracting the known angles of
triangle D from one hundred and eighty degrees,
180°. Here, 180° - 90° -19.65382406° =
70.34617594.

   Next, let us observe the values of all of the
angles of this rectangle.

Angles

01   23.19738752°
02   67.16634583°
03   22.83365418°
04
05   19.65382406°
06   70.354617594°
07   66.8261248°
08
09
10   90.0°
11   90.0°
12   90.0°

This leaves angles 04, 08, and 09 or the angles of triangle A.

To find the value of angle 9, we should observe that a straight line has 180 degrees in its angle. This is because one side of a right triangle was placed down so that it points in the same direction as another line of that right triangle. Now, because there are only two lines that point in the same direction, there is no longer a triangle, but we can see that the angle of the straight line is 180 degrees or two times 90 degrees. The angles of 1, 6, and 9 make up a straight line that exist between angles 10 and 11. We know the values of Angle 1 and Angle 6, which are 23.2 and 70.35, respectively. So

$180° - 23.19738752° + 70.34617594°) = 86.45643654°$
for angle nine, 9.

The same thing can be done to find the value of Angle 8. Here, it plus the value of angle two, or 67.16634583°, plus the value of Angle 7, or 66.8261248, is

$180° - 67.16634583° + 66.8261248°) = 46.04752937°$

To find the value of angle 4, notice that we are not dealing with the total value of the angle of a straight line but the total value of a right angle. This is because three parts of the rectangle are of this top left side of the rectangle. Two parts of the right triangle have the values of 22.83365418 degrees and 19.65 degrees for Angle 3 and Angle 5, respectively. Thus,

$$90° - (22.83365418 + 19.65382406)° = 47.51252176°.$$

Here are the values of the angles.

Angles

```
01  23.19738752°
02  67.16634583°
03  22.83365418°
04  47.51252176°
05  19.65382406°
06  70.34617594°
07  66.82612480°
08  46.00752937°
09  86.45643654°
10  90.00000000°
11  90.00000000°
12  90.00000000°
```
_____

    720.00000000°

Because the Right Angles, Angles 10 = 90°,
                        Angles 11 = 90°,
                        Angles 12 = 90°,
                                    and
                Angles 3, 4, 5 = 90°,
plus the Straight Line Angles of Angles 1, 6, and 9 equal   180°
                                    and
    Straight Line Angles of Angles 2, 7, 8 = 180°,
_____
    They all sum to   720°.

In summary, what we have observed is that the sides of triangle A can be found by using either the Law of Sines or the Pythagorean Theorem. Also, we could have used the Law of Cosines to find the answer of the values of the sides of triangle A.

Here, is the Law of Sines.
a/Sin A = b/Sin B = c/Sin C

Here, is the Law of Cosines.

$A^2 = b^2 + c^2 -2bc \cos A$

$B^2 = a^2 + c^2 -2ac \cos B$

$C^2 = a^2 + b^2 -2ab \cos C$

After obtaining the values of the sides of Triangle A, we can find the value of the size of Triangle A by using **Heron's Formula**.

Here SS represent "Sum of the Sides."

$$((SS)/2 \ [(SS)/2 - a] \ [(SS)/2 - b] \ [(SS)/2 - c])^{0.5}$$

```
Side a = 59.46427499
Side b = 82.46211251
Side c = 60.92618485
--------------
        202.852574
(SS)/2 = 101.4262862
```
So in Heron's formula we have

(101.4262862) times (101.4262862-59.4627499) times (101.4262862-82.46211251) and times (101.4262862-60.92618485)$^{0.5}$

$=(101.4262862)(41.96201121)(18.96417396)(40.50010135)^{0.5}$

$= 1808.000015$, or the area of triangle A.

Previously we arrived at the value of triangle A to be about 1808. Here, the found answer is 0.000015 more than the first answer.

Next,

again, the value of all four triangles should sum up to the value of the size of the big rectangle. The area of this rectangle is 4256, or 76(56).
Subtracting triangles B, C, and D, respectively, from this should give the value of Triangle A.

```
      4256.00
-   1,216.00
-     672.00
-     560.00
   -----------
    1,808.00
```

Close, and good enough.

Let us see you find the values of the sides of triangle A by using the Law of cosines.

Cuban Link

In the name of the Father, Son, Holy Spirit Amen
Please forgive me Lord I know I'm misbehaving
I'm staying up at night just blazin
Thinking about my life and this ---- -- situation
Satan's waiting patient with his temptation
Trying to make his way so he can take control
I know I God's creation I won't sell my soul I know my foundation
Show me the road where I go right or left love or hate life or death
Am I bait for the snake only fate knows the rest
I got questions - yeah– Oh Lord I got questions

Excuse me Father – can I get a little bit of your time
I don't mean to bother but I got a lot of things on my mind
See I got these problems and I don't want to go tot my nine
And I'm trying to solve them but it feels like I'm running out of time
running out of time - so I call upon ya - so I call upon ya
I call upon ya - I call upon you

Pardon me Lord It's kinda hard for me part open these doors
It's got to be more to life than just parties and broads
My mind is so lost although my heart is guided by yours
I crossed the road and ended up where I started before
From Poor to Entrepreneur performing raw without an album in stores
doin tours from Cali down to Harlem N.Y.
Problem of all sorts can't dodge 'em keep getting caught
I thought the art of war was stronger than the arm of the law
I caught a felony and though I got locked up before
It all fell on me all without probable cause
I heard you telling me jealousy's a part of this sport
I felt your energy like Lazarus I'm guarded by dogs
Who though I'd be the one the audience applaud
Who thought I'd get to see my face in the Source
Who'd thought I'd be double crossed
I know God was the force that kept my soul strong threw it all
It's too far to walk the dog and throw it all out the door
Help me Lord!!

Excuse me Father – can I get a little bit of your time
I don't mean to bother but I got a lot of things on my mind
See I got these problems and I don't want to go tot my nine
And I'm trying to solve them but it feels like I'm running out of time
running out of time - so I call upon you - so I call upon you
I call upon you - I call upon you

Now I lay me down to sleep I pray my lord my soul to keep
And if I die before I wake I pray my lord my soul you take
I'm just a man I make mistakes learn to separate the real from the fake
Gotta keep the faith by praying everyday
Shine you light on me Lord before it's too late

Excuse me Father – can I get a little bit of your time
I don't mean to bother but I got a lot of things on my mind
See I got these problems and I don't want to go tot my nine
And I'm trying to solve them but it feels like I'm running out of time
running out of time   Excuse me Father (repeat)

Part Fourteen Problems

It is wise to do these problems without use of a calculator.

14A.) Seven numbers differ by twenty-eight in value, where the largest number is thirteen and one-half. What is the sum of this series of numbers?

14B.) $(X + 2Y)^2 = 100$, when 0 is the value of X what is the value of Y?

14C.) What is the radius of a circle that has a circumference of twenty-two?

14D.) $12!(66)^2 = ?$

14E.) $X^2 - X + 3 = 0$, solve for X.

14F.) $2Y^2 - 2Y + 8 = -2$, Solve for Y.

14G.) Six numbers are summed and have a difference of eight between them. The summed value of them is eight. List these six numbers.

14H.) List the law of Sines.

14I.) Write 0.116 as a percent.

14J.) A triangle has corner locations of (-2, -19), (4, -1), and (6, 5). What is
      the area of this triangle?

A New York Train

Part Fifteen.
Comic Relief

With this story hear a jazzy-funk chanting and repeating sound with instruments and voice creating a "P-funk" sound.

<u>Comic Relief</u>

As a preteen, he noticed that he could jump in to the air and float down to the ground. Nobody else in his family could do this. His friends loved this and often made laughter with him as he appeased what laughter they wanted to give. At times he bought this characteristic to the basketball court. His friends labeled him "Comic Relief" because of his jump and float basketball technique.

Unlike his older brother, he did not have to chose between work and school, so he chose to go to college. He was sixteen years younger than Mti and with his family's gains in income his father, who was born in just north of Dar es Salaam, Tanzania but grew up in Kampala, Uganda, with his mother, who was born in Fort Lauderdale, Florida, decided to give him an European first name, Daniel, and an American way of life.

Like his brother and father he grew to be tall and, like his brother, he did very well in school at a local Catholic high school. Still, his interests were in turning into a professional basketball player. Both of his parents thought this to be a waste of time. Yet, they often attended his high school games where he averaged twenty points a game. His parents wanted to know what are you going to do for a living. He certainly did not want factory or clerical work. He hated cab driving, which his father often did.

For the most part, he liked to party, drink, date, and play basketball. Looking at his aged parents and older brother did not give the working life a good start. It was offered to him by his brother's work place but he refused and studied so he could go to college on an academic scholarship, which he later received soon after graduating from high school.

Keeping his grades above "party time demands" proved to be trouble. Just while he was ready to flunk out he jumped twice into the air during a basketball game and distracted the opposition's attention. As a result, his team was able to regain the lead in both games against different teams. He had only tried this trick of jumping and floating four times in college before then and twice in a row this technique worked well. With that, his grade point average and his love life excelled.

In numerous following games he played well but did not use his jump and float move. The college team he played for advanced to the finals. In the championship game he was expected to be one of the most important, not the most important, player for this team. The opposition was faster, taller, and heavier than his team members. Yet, neither team ever maintained a lead greater than five points for two minutes in the championship game. Over time, yes overtime, was expected by news reporters weeks before the play-offs between these teams began.

With two minutes left and behind by two points his team put the pressure on the floor by accelerating the pace of the ball. The opposition's lead was increased to three points before Daniel closed it down to only being a one point lead. Not winning, the game became a test against "time." The opposition was able to accelerate with the pressure. So, Daniel decided to mix-up the pace by jumping and floating to the ground. This worked to distract the opposition's attention intensity. But it also caused one of his team members to throw the ball at him while he was floating down to the ground. The ball hit his arms and he could not grab and keep it. It went to various parts of the court before being used in a slam to give the opposition another three point lead that was increased with a related fowl shoot. His team returned after a "time-out" to tie the game but could not maintain this and lost the game by one point.

Almost everyone, including news reporters, blamed the lost on Comic Relief and his jump and float technique. Although his girlfriend, Shirl, stayed behind him his school grades showed low grades.

He dropped out of school and found time to party and supported himself by selling marijuana mostly and rarely some other illegal drugs. One night while with friends in an apartment, two other drug dealers decided to go to the nearby 7-11 store. In the store, a man and a woman noticed the two dealers. The man was a son of a police officer named Crystal Critter. He, the man, saw the two dealers enter the store and pursued them for drugs so that he with his girlfriend could use the drugs to get high. The two dealers knew him, his girlfriend, and the man's mother, Officer Critter. The clerk behind the cashier noticed the transaction and called police who were only one block away. With no drugs on him, Daniel left the apartment to go to the 7-11 store. He arrived five seconds before police arrived and he spoke to his friends. Then the police arrived and searched everyone he knew that was in the line of customers.

The police found no drugs on the lady but found some drugs on the man who rented the apartment where Daniel left his drugs and some of his clothing. A small knife was found on the woman. So, the police arrested all five of the suspects.

In court he did not get a good defense and was sent to jail for thirteen months. His time in the "slammer" was almost eventless and his release quickly arrived. Upon his release, his brother's job approached him and this time he accepted their work offer, which not only pleased both his parents and his brother but gave his life some practical direction. Moreover, the company skipped the practice of making him work four weeks before getting his first bi-weekly paycheck and gave him his first check after two weeks of full-time work.

He went to the bank, United Bank of Pemberton, which housed his employers account to cash his check. He did not have to pay a fee to cash his check.

In the bank, a mother and a little girl entered. The mother quickly got in the bank's line of customers. Her pre-school daughter seemed to be bored with the bank, so he tried to uplift her spirits by asking her "What does she want to do when she grows up." He found out that she did not want to play basketball, be a fire man, or work in the "town's company-store." She wanted to work in a circus. They talked about training for that type of activity and how to train animals of the circus. She told him that "People aren't the only animals in a circus" and he agreed. When her mother was next to speak with the teller, he decided to show her his circus act of jumping and floating to the ground. She was amazed and delighted at his work on the bank floor.

At that time four people, including a woman and two men with guns, entered the bank. The unarmed man was the son of the police officer he knew as Officer Critter. One of the gun men told people in the bank to remain calm. Later, he noticed Comic Relief's small jump into the air and before floating down to the ground, then he shouted "I told you ta be quiet!" He shot at Daniel four times, hitting him twice in the torso and once in the right leg.

The men and woman escaped without taking more than a handfull of dollars from one of the bank tellers. One of the men's mother and other police officers arrived and took Daniel to a hospital. His chances of a good recovery were small. His basketball school insurance had expired and his hospitalization bill was very large. To help him various groups, who knew of his basketball playing a few years earlier, created a fund for his recovery. The second largest donator to the fund was his brother's sports club. Their donation was substantial. This with hospitalization and prayer caused him to recover much and fully.

He decided not to return to basketball playing even though a professional team did consider working with him. Five months later, Shirl proposed to him for marriage after he accepted work at his brother's former job.

Write your opinions about this story if extra paper is needed feel free to use it.

This is a picture of an Odunde Celebration of Philadelphia, PA and a Housing Project of Newark, New Jersey

FIRE
ANT"

In 1997 I met a cab driver in Montgomery, Alabama, after I walked and jogged from Selma to Montgomery, Alabama along Route 80, the same route that people walked to get voters' right in 1963. It is a forty-five to fifty mile hike. The cab driver told me that his brother was killed by a group of "Fire Ants." His brother got drunk and fell out on some steps. The fire ants found him and started to eat him. Some of their chemicals can cause an animal to become paralyzed. In that condition the ants were able to kill his brother.

Half – Life   The natural log, ln, is based on the number

being raised to various exponential powers, e^x. Here, e to the first power is 2.718281828. This can be found by pressing one on your calculator before hitting the shift key and then the e^x key. Here, you've just raised e to the first power, where anything raised to the first power equals its self. Another way to find the value of e is to sum one plus another one which is divided by a large number. After summing this raise the sum to a power that is equal to the above large number.

   Example: $[1 + 1/(A\ large\ number)]^{(A\ large\ number)} = [1 + 1/1{,}000]^{1{,}000} = 2.716923932$
A closer result to the value of e can be made by dramatically increasing the value of the large number.   Example: $[1 + 1/1{,}000{,}000{,}000)^{1{,}000{,}000{,}000} = 2.718281828$

This is very close to the true value of e.

J92340u52349uv09358u2904u92u4904u294292904u52391u5v-4u2-up9-0u9-q2u94u294

Mysteries of Great Zimbabwe

- By Peter Tyson
- Posted 02.22.00
- NOVA

The first whispered reports of a fabulous stone palace in the heart of southern Africa began dribbling into the coastal trading ports of Mozambique in the 16th century. In his 1552 *Da Asia*, the most complete chronicle of the Portuguese conquests, João de Barros wrote of "a square fortress, masonry within and without, built of stones of marvelous size, and there appears to be no mortar joining them."

De Barros thought the edifice, which he never saw, was Axuma, one of the cities of the Queen of Sheba. Other Portuguese chroniclers of the day linked the rumored fortress with the region's gold trade and decided it must be the biblical Ophir, from which the Queen of Sheba procured gold for the Temple of Solomon.

The view from Great Zimbabwe, which Europeans didn't "discover" until the 1870s Enlarge Photo credit: © Karen Graham/iStockphoto

A city of stone

147

This notion persisted for centuries, right up until the monument's 19th-century European "discovery." That distinction fell to a young German named Carl Mauch. In 1871, Mauch, eager to seek for the fabled ruins of Ophir, penetrated deep into what is today southern Zimbabwe. In August, he reached the home of a lone German trader, who told him of "quite large ruins which could never have been built by Blacks." On September 5, local Karanga tribesmen led Mauch to the site.

In the midst of a wooded savanna backed by bare granite hills stood a city of stone. Its beautifully coursed walls curved and undulated sinuously over the landscape, blending into the boulder-strewn terrain as if having arisen there naturally. Bearing no mortar, as de Barros had correctly heard, the walls nevertheless reached enormous height, standing as high as 32 feet over the surrounding savanna. Of fully 100 acres of these granite enclosures, not a single one was straight.

Mauch was looking at the greatest pre-Portuguese ruins of sub-Saharan Africa.

Inside the Great Enclosure at Great Zimbabwe Enlarge

Dim view

Unfortunately, Mauch, for all his tenacity, was "no thinker," as Peter Garlake, author of the definitive archeological text on Great Zimbabwe, deemed him. And Mauch only boosted the Portuguese theories of three centuries before. The soapstone and iron relics he uncovered told him that a "civilized [read: white] nation must once have lived there." From a lintel, he cut some wood that he described as reddish, scented, and very like the wood of his pencil. Therefore, he concluded, the wood must be cedar from Lebanon and must have been brought by Phoenicians. And *therefore,* the Great Enclosure—the edifice's most impressive structure, which local Karanga called *Mumbahuru,* "the house of the great woman"— must have been built by the Queen of Sheba.

Mauch was looking at the greatest pre-Portuguese ruins of sub-Saharan Africa.

As it turns out, Mauch's description of the wood aptly characterizes the African sandalwood, a local hardwood that later visitors also found in the walls of the Great Enclosure. But no one would know that for years.

In the meantime, Mauch's line of reasoning, distinguished as it was by the most purblind logic, perfectly suited Cecil Rhodes, whose British South Africa Company (BSA) occupied Mashonaland in 1890. (Mashonaland lies just to the north of Great Zimbabwe.) Inextricably steeped in his native country's racist views, Rhodes bought into Mauch's take without a second thought. Indeed, on Rhodes' first visit to the site, local Karanga chiefs were told that "the Great Master" had come to see "the ancient temple which once upon a time belonged to white men."

The highest of Great Zimbabwe's walls stand three stories tall. Enlarge Photo credit: Cicada Films

Ignoring the obvious

Eager to nail down the edifice's exotic origins once and for all, Rhodes and his BSA quickly sponsored an investigation of Great Zimbabwe. They hired one J. Theodore Bent, whose only claim to expertise lay in an antiquarian interest born of travels through the eastern Mediterranean and Persian Gulf. Bent adhered just as tenaciously as Rhodes to the notion of the city's non-black origin, though to his credit he didn't automatically swallow the link to the Queen of Sheba. (As he set to work at Great Zimbabwe, he later recalled, "the names of King Solomon and the Queen of Sheba were on everybody's lips, and have become so distasteful to us that we never expect to hear them again without an involuntary shudder.")

All the artifacts Bent subsequently uncovered screamed "indigenous." Pottery sherds and spindle whorls; spearheads of iron, bronze, and copper; axes, adzes, and hoes; and gold-working equipment such as tuyères and crucibles—all were very similar to household objects used by the local Karanga. Yet Bent, clearly incapable of following where the evidence might lead him, concluded ("a little lamely and nebulously," notes Garlake) that "a prehistoric race built the ruins ... a northern race coming from Arabia ... closely akin to the Phoenician and Egyptian ... and eventually developing into the more civilized races of the ancient world."

149

All artifacts that Bent turned up pointed to an indigenous origin of Great Zimbabwe and its builders, but he would have none of it. Enlarge

"Reckless blundering"

Bent was amateurish and narrow-minded but not utterly incompetent. The same could not be said of Richard Nicklin Hall, a local journalist and author of *The Ancient Ruins of Rhodesia*. In what would prove to be one of the most sickeningly misguided assignments in the history of archeological preservation, the BSA appointed Hall Curator of Great Zimbabwe, with a mandate to undertake "not scientific research but the preservation of the building."

Instead, Hall, hell-bent on finally settling the issue of its origins, launched into a full-scale "archeological" investigation. Claiming he was removing the "filth and decadence of the Kaffir occupation," he scoured the site for signs of its white builders, discarding from three to 12 feet of stratified archeological deposits throughout Great Zimbabwe. An archeologist who visited the site shortly after Hall left deemed his fieldwork "reckless blundering ... worse than anything I have ever seen."

Great Zimbabwe was home in its heyday to some 12,000 to 20,000 people.

Word eventually got back to the BSA of Hall's desecration of southern Africa's greatest archeological treasure, and he was dismissed. But the damage had been done. "Hall's disastrous activities left only vestiges of archeological deposits within the walls," wrote Garlake in his book *Great Zimbabwe*, "a paucity that was to inhibit all future scientific work."

David Randall-MacIver, the first archeologist to study Great Zimbabwe, declared it unequivocally of African origin.

Archeology begins

Contrite, the BSA hired archeologist David Randall-MacIver, protégé of the great Egyptologist Flinders Petrie, to investigate the site. Hall's polar opposite in almost every way, Randall-MacIver quickly concluded that former mud dwellings within the stone enclosures "are unquestionably African in every detail and belong to a period which is fixed by foreign imports as, in general, medieval."

While MacIver's careful work set the stage for the sound archeological inquiry of Great Zimbabwe, racial prejudice surrounded the monument until quite recently. In the 1960s and 1970s, as the edifice grew into a potent symbol of the African Nationalist movement, the white government of Rhodesia set about suppressing the findings of prehistorians who claimed that Africans had built Great Zimbabwe. (Garlake, for one, was forced out of the country.) But those problems went away when Zimbabwe, as the country is known today, achieved majority rule two decades ago, and now we can look at Great Zimbabwe free of racial overtones.

"Venerated houses"

Many believe that "Zimbabwe" is a contraction of the Shona phrase *dzimba dza mabwe*, "houses of stone." (The Shona are Bantu people of Zimbabwe and southern Mozambique.) Garlake, for his part, feels the word more likely derives from *dzimba woye*, "venerated houses," a term usually reserved for chiefs' houses or graves.

Either way, archeological investigation has shown that the edifice's monumental walls did once enclose houses. Great Zimbabwe was a city, home in its heyday to some 12,000 to 20,000 people. To this day, *daga*, a clayey conglomerate of gravel that is Africa's most common indigenous building material, still stains the soil within Great Zimbabwe a robust red color.

While few traces of the mud houses remain, the towering stone walls stand in mute testimony to the city's former greatness. Quarried from the nearby granite hills, the rock used in the walls' construction easily split along fracture planes, giving the stones a cuboidal shape that lent itself to stacking without need of mortar. Ranging from four to 17 feet thick, Great Zimbabwe's walls are about twice as high as they are wide. This results in a very sturdy structure, which spreads its pressure evenly over the ground and adjusts well to subsidence. When two walls meet, they abut eachother with unbroken vertical joints; there are no interlocking stones. In the finest walls, workers knapped and dressed the stones so well that the coursing is as smooth as a modern brick wall.

Whites did not build Great Zimbabwe, Blacks did, and this fact only deepens the sense of mystery.

The Great Enclosure is the largest single prehistoric structure south of the Sahara. Looking from the air like a giant gray bracelet, its elliptical Outer Wall is more than 800 feet long and contains an estimated 182,000 cubic feet of stone, more than in all the site's other ruins combined. Garlake believes the Great Enclosure, which encircles a series of smaller stone walls and a Conical Tower shaped like a stone beehive, was likely a royal residence.

Archeologists have determined that the Conical Tower is completely solid. Its original purpose remains unknown.
EnlargePhoto credit: Cicada Films

Rise and fall

While the site was occupied in ancient times—iron was in use there by the third century A.D.—its rise to prominence, and the advent of the finest walls, occurred in the 14th and 15th centuries during a florescence in trade. Great Zimbabwe happened to lie right on the route between the region's gold-producing regions and ports such as Sofala on the Mozambique coast, where merchants traded African gold and ivory for beads, cloth, and other goods from Arabia and farther east. The site may also have been a religious center, as evidenced by stone monoliths and "altars" found throughout the site, along with enigmatic soapstone birds and figures that, says Garlake, "point to the important role of ritual and symbol in the art and architecture of Great Zimbabwe."

By the mid-15th century, however, the balance of trade had shifted to the north. Local resources had also apparently dwindled to dangerously low levels from overuse, and salt was scarce. Whatever the cause, Great Zimbabwe's people abandoned their once-glorious stone city, leaving the site a ruin that Mauch found 400 years later inhabited by local Karanga people who had no idea of its history.

The Lemba, including their spiritual leader Professor Matshaya Mathiva (seen here), believe their ancestors erected Great Zimbabwe. Enlarge Photo credit: Cicada Films

Unsolved mysteries

Despite decades of study, mysteries still cling to Great Zimbabwe like ivy. How did its residents manage to monopolize trade in the area? To what degree was it a religious center? Why was it abandoned? Even the question that, as Garlake said about the site itself, "has given rise to such strong, widespread, and often bizarre emotional responses"—who built it?—has been only partly answered.

To wit: *Which* Africans built it? Many tribes, including the Shona and Venda, maintain that their ancestors were responsible for Great Zimbabwe, but the Lemba are "particularly insistent," says University of London scholar Tudor Parfitt. "They claim that one of their clans, the Tovakare, were the actual builders of Zimbabwe," Parfitt says. "They even call them Tovakare Muzimbabwe, which means 'the ones that built Zimbabwe.'"

Certain evidence appears to support the Lemba claim. For instance, unlike other Bantu tribes, who bury their dead in a crouched posture, the Lemba bury theirs in an extended position, as did the ancient Zimbabweans. One of the strongest pieces of evidence concerns trade, Parfitt says. "Great Zimbabwe was a civilization that was constructed very largely on wealth generated from cattle and trade. And given that for hundreds of years we know the Lemba were the great traders of southern Africa, it seems almost certain that their ancestors would have been involved in this trading nexus between Great Zimbabwe and the Indian Ocean."

"The mystery of Zimbabwe is the mystery which lies in the still pulsating heart of native Africa."

153

If the Lemba contention is true, does this mean that outsiders—that is, not native Africans—built Great Zimbabwe? After all, the Lemba have Semitic origins (see Tudor Parfitt's Remarkable Quest). The answer is no, because by the time Great Zimbabwe was built in medieval times, the Lemba had become decidedly African, having so thoroughly intermixed with Bantu Africans over many hundreds of years that today, among other African traits, the Lemba have dark skin and speak a Bantu language.

Indeed, the more contentious part of that question "who built it?" has finally been put to rest almost 450 years after João de Barros and others first propounded it. Whites did not build Great Zimbabwe, Blacks did, and this fact only deepens the sense of mystery enveloping the site. As archeologist Gertrude Caton-Thompson declared back in 1931:

Examination of all the existing evidence, gathered from every quarter, still can produce not one single item that is not in accordance with the claim of Bantu origin and medieval date. The interest in Zimbabwe and the allied ruins should, on this account, to all educated people be enhanced a hundred-fold; it enriches, not impoverishes, our wonderment at their remarkable achievement ... for the mystery of Zimbabwe is the mystery which lies in the still pulsating heart of Native Africa.

Questions

47.) Of the following graphed triangle find various elements of it.

Find the degree of each of the angles.

a.) Angle A =____ b.) Angle B =____ c.) Angle C = _____

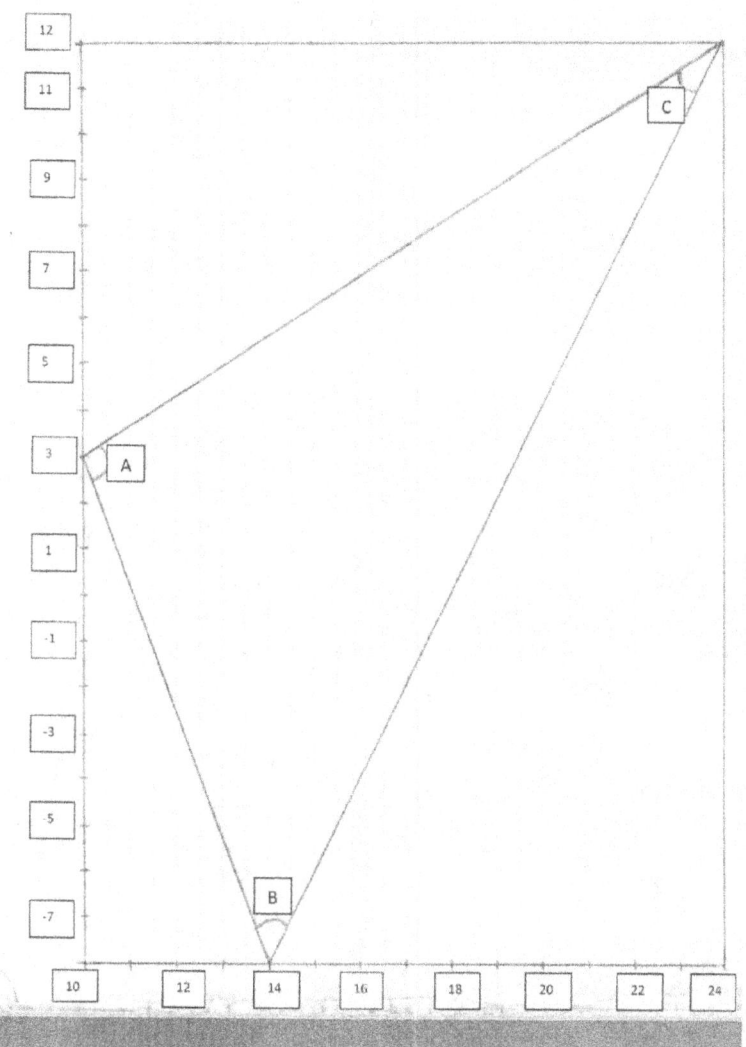

156

48.)   Find the lengths of each of the sides of the triangle.

49.)   Find the area of the triangle.
       Heron's formula is $[(SS/2)(SS/2-a)(SS/2-b)(SS/2-c)]^{0.5}$,
       where SS represents the sum of the sides of the
       triangle.

50.)   Ln of 45/30 = ?

51.)   In a graph where are the two places where the function
                3X meets the other function $16X^2-9X-4$?

52.)   What is the distance between points 66X, 3Y and 15X, -18Y?

53.)   Find the distance between the circumferences of the two
       following circles. The top circle has a radius of
       5.852349955 yards and the inside triangle reaches the,
       within a circle, x axis at zero and at 5.5. Its origin is
       at(7, -16). The bottom circle has a radius of 5.656854249
       and the inside triangle reaches the x axis at zero and
       four. Its origin is at (-10, 8.5). Also, what is the
       purpose of reporting the coordinates of the triangles that
       are inside of the circles?

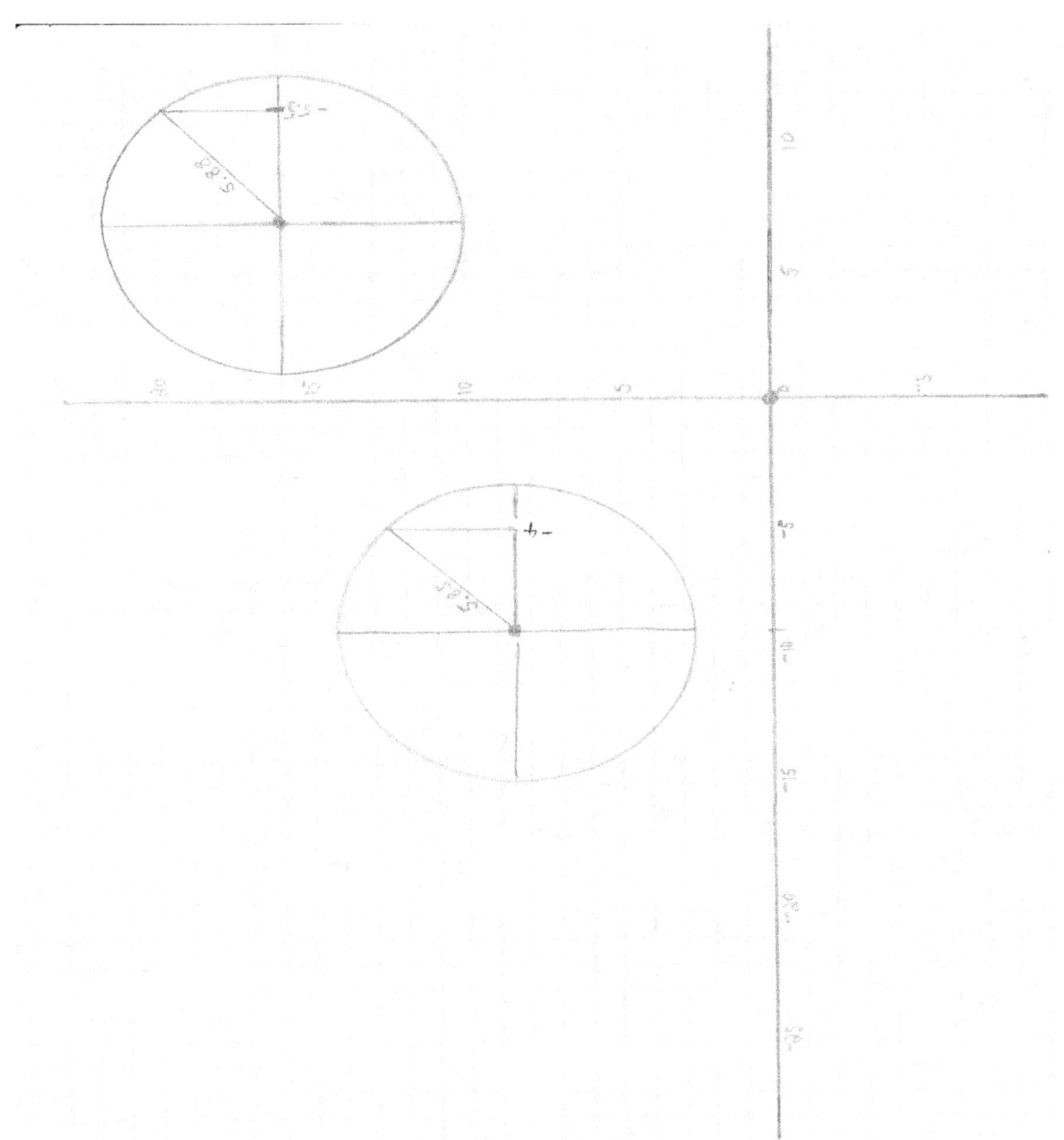

        In the above graph the vertical line is the X axis
and the horizontal line is the Y axis.

54.) The bottom left side angle of the above triangle is 52.59464337°, the cosine of this angle times what equals the length of the triangle's X axis and the sine of this angle times what equals the length of the triangle's Y axis.

Again, the bottom left angle of the above triangle is 52.59464337°. What times the bottom left angle's degrees equals the length of the triangle's X axis and what times the bottom left angle's degrees equals the length of the triangle's Y axis? The left side corner of the triangle is at (-6.5, -1.5), the bottom right corner of the triangle has a location of (6.5, - 1.5), and the top right location of this triangle has a location of (6.5, 15.5.)

The bottom left side angle of the above triangle is 52.59464337°, the cosine of this angle times what equals the length of the triangle's X axis and the sine of this angle times what equals the length of the triangle's Y axis.

Again, the bottom left angle of the above triangle is 52.59464337°. What times the bottom left angle's degrees equals the length of the triangle's X axis and what times the bottom left angle' degrees equals the length of the triangle's Y axis? The left side corner of the triangle is at (-6.5, -1.5), the bottom right corner of the triangle has a location of 6.5, -1.5), and the top right location of this triangle has a location of (6.5, 15.5).

The Y axis

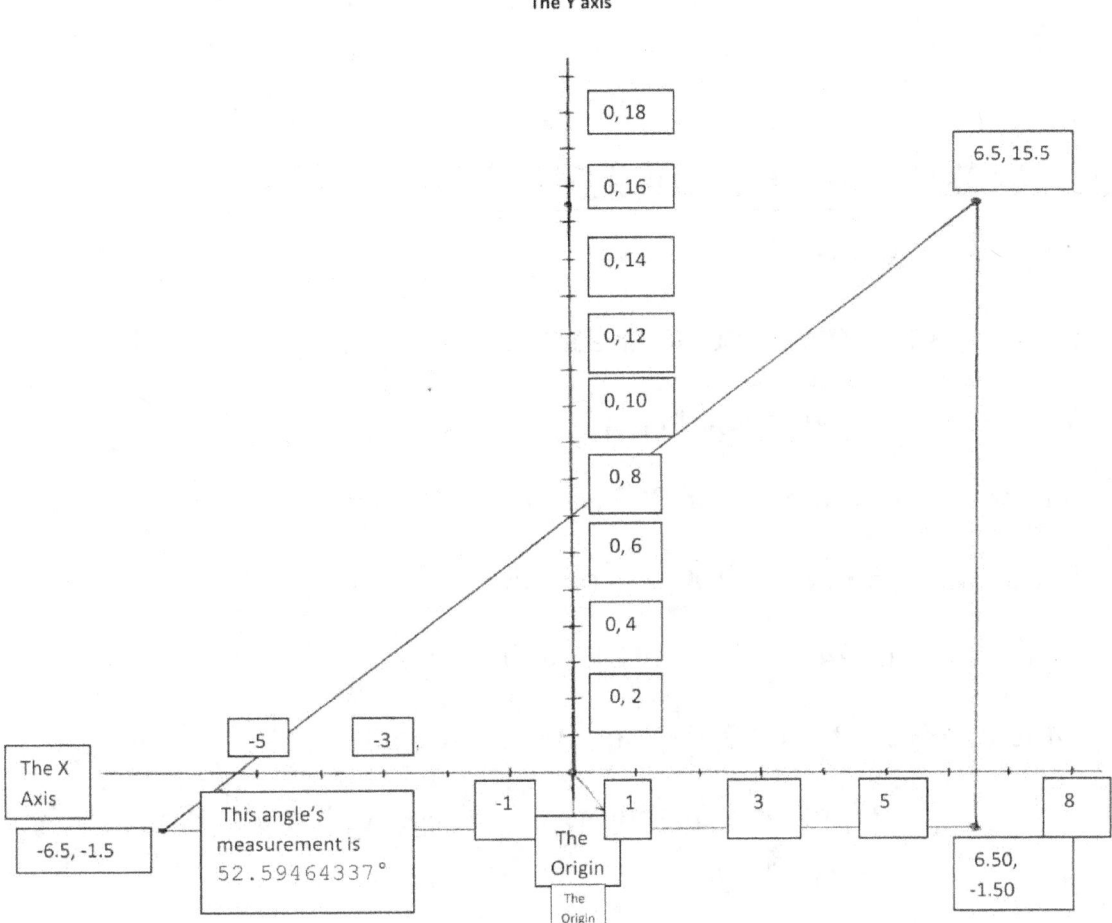

0, 18

6.5, 15.5

0, 16

0, 14

0, 12

0, 10

0, 8

0, 6

0, 4

0, 2

-5   -3

The X
Axis

-1

1   3   5   8

-6.5, -1.5

This angle's
measurement is
52.59464337°

The
Origin

The
Origin

6.50,
-1.50

55.) $16X^2 - 9X - 4 = 3X$, solve for X.

56.)

Crime in Autaugaville, Alabama

In 2001 there were

- 0 murders or 0.0 per 100,000 people

- 1 rape or 122.0 per 100,000 people

- 0 robberies or 0.0 per 100, 000 people

- 5 assaults or 609.8 per 100,000 people

- 2 burglaries or 243.9 per 100,000 people and

- 14 larceny counts or 1,707.3 per 100,000 people.

Population Questions   Do not use a calculator to do these questions.

1.) With the above information tell us how many people lived in this town in the year 2001.

2.) Imaginationville, Pennsylvania, a fictional town, had an eight and one-half percent yearly birth rate of its 2009 population of 18,902 people. Find the number of births per 100,000 people of Imaginationville of that year.

Part Fifteen Problems

(Do all multiplication, division, addition, and subtraction without use of a calculator.)

15A.) $-608 -808^2 = ?$

15B.) Where do $5X^2 - 8 + 3$ and $2X^2 - X + 9$ meet?

15C.) If a population of 61 people grew by six percent for twelve years. How many people are in that population after ten years if the population had an equal amount of deaths as it had births and no one left from or arrived to the area of this population?

15D.) $66^3 - 66 =$ _____?

15E.) By January 1, 2011, South Sun, Florida, a fictional town, had an twenty and one-quarter percent yearly birth rate of its 1997 population of 8,300 people. Find the number of births per 100,000 people of South Sun, Florida of that time period.

15F.) $22X^2 + X = 9$

15G.) List Heron's Formula.

15H.) $2X^2 + X = 9$, What is the value of X?

15I.) $2X^2 + 13$ meets $5X$ where?

15J.) A population of 66,000 bacteria decreases by twelve percent every day. When will it be half gone?

Part Sixteen Log and Ln Facts

How many powers must be raised for it to reach a particular value can be used and explained with an exercise called the "Half-Life." A particular strand of a virus creates mild discomforts in human beings for a time period of up to three weeks. Applying good rest and meals with mild medication will stop the virus and reduce its population at the rate of one-half of its population size every 2.23 days. A particular person was infected by 380,083 of organisms of this virus. She used prescribed mild medication, ate well, and rested, well. How many days must go by before the virus only has a population that is at the safe level of 98,000 organisms?

Log [(A)/(B)] = Log A-Log B

Log[(A)(B)] = Log A+Log B

Log[(A)(A)(A)] =

Log A + Log A + Log A = 3Log(A)

Log(8)   =Log(10)$^{0.903089987}$

0.903089987Log(10) = 0.903089987
and 10$^{0.903089987}$ = 8.

Ln[(A)/(B)] = Ln A - Ln B

Ln[(A)(B)] = Ln A + Ln B

Ln[(A)(A)(A)] =

Ln A + Ln A+ Ln A   = 3Ln(A)

Ln (8) = Ln (2)^3 = 3Ln (2)
= 2.079441542 and e$^{2.079441542}$ = 8.

Here, to solve this problem, we must determine the number of half-lives that will cause the number 380,083 to be equal to 98,000. $380,083/2^X = 98,000$.
So, $380,083 = 98,000(2^X)$. Solving for X shows $380,083/98,000 = 2^X$

To find the power that 2 must be raised in order for it to equal 380,083/98,000 use
Ln (380,083/98,000) = ln2^X. This will equal ln (3.878397959) = ln2^X.
Next, 1.355422171 / ln 2 = X

1.355422171 /0.69314718 = X = 1.9554660846

This answer must be multiplied by the number of days, 2.23, that the treatment takes to reduce the virus population size to being one-half of this size at a particular time.
(1.9554660846)(every 2.23 days) = 4.360677687, or 4.361 days.

If you forget the value of e, remember that   [1 + 1/(a very large number)]$^{(The same very large number)}$= e or the true e, 2.718281828.

Example: $[1+1/(1,000,000)]^{1,000,000}$ = 2.718281693 which is about the true value of e.

Part Sixteen Problems

16A.) A series of seven numbers has a difference of -0.9 between the elements and the elements of this series sum to a value of twenty-two. List the seven numbers.

16B.) $(-4X + 3)^2/2X^2 = 6$ Solve for X.

16C.) Ln 25/Ln20   = What?

16D.) A group of eighteen numbers differ in value by one-half and sum to a value of six hundred and fifty, 650. What is the value of the smallest of these numbers?

16E.) $e^x = 172,000$, solve for X.

16F.) How many equations must one have to solve a three equation matrix?

16G.) $(27/9^{-3}) + 99 = X + 10$. Solve for X.

16H.) $\underline{\text{Ln } 180 + \text{Ln } 6}$ = What?
         Ln 3

16I.) $(2X - 33)^2 + 8.81 = 88$, solve for X.

16J.) Triangle A has an angle that is measured at 72 degrees at point C and 41 degrees at Point A. Side b was measured and found to be 26.14634146 inches long and side c was found to be 27.01414171 inches long. Find the angle at Point B and the length of triangle side a by using both the Law of Sines and by using the Law of Cosines.

Two Pictures of Newark, New Jersey

Part Seventeen

A.) Place the divisor and the dividend in order according to the powers of the unknown and where there is no element for a particular exponent place a zero times this exponential value of the unknown.

B.) Use the first term of the divisor to show how many times it can go into the first term of the dividend.

C.) Multiply the answer by each element member of the divisor and place the results in appropriate places of the dividend.

D.) When the divisor cannot go into any of the resulting figures without resulting in a negative exponent value   then the numbers there are the remainder of the answer that will become the numerator of a fraction and the denominator of that same fraction of the answer.   **Here is an E X A M P L E.**

$$\frac{6X^4 + 4X - 3}{3X^2 - 2X + 1} \quad = \quad \frac{6X^4 + 0X^3 + 0X^2 + 4X - 3}{3X^2 - 2X + 1} \quad =$$

$$2X^2 + 1.3333X + 0.2222$$

$3X^2 - 2X + 1 \big\backslash\ 6X^4 + 0X^3 + 0X^2 \quad\quad + 4X\ -\ \ 3$

$\underline{-(6X^4 - 4X^3 + 2X^2 \quad\quad\quad)\quad ._____}$

$\quad\quad\quad 4X^3 - 2X^2 \quad\quad + \quad 4X$

$\quad\quad\quad \underline{-(4X^3 - 2.6666X^2 + 1.3333X)}$

$\quad\quad\quad\quad\quad 0.6666X^2 + 2.6666X - \quad 3$

$\quad\quad\quad\quad\quad \underline{-(0.6666X^2 - 0.4444X + \quad 0.2222)}$

$\quad\quad\quad\quad\quad\quad\quad 3.1111X - 3.2222$

So, the answer is $2X^2 + 1.3333X + 0.2222$ and the remainder is $\dfrac{3.1111X - 3.2222}{3X^2 - 2X + 1}$

Should you multiply this answer by the divisor you will reach the number that is the numerator of this fraction.

Part Seventeen Problems

17A.) Factor $16X^2 - (3X)^2$.

17B.) List the Law of cosines.

17C.) How many cubic feet are in a cube with sides that are two meters long?

17D.) $\dfrac{18X^{10} - 2X}{X} = 0$ Solve for X.

17E.) <u>Solve the following matrix.</u>
$$-6X + 0.2Y \ 08C + 9 = 9$$
$$-C + 10Y - 6X = 22.5$$
$$-10Y + 2X + 6C + 6 = -16$$

17F.) $(21X^2 - 4X - 4)/12 = 18$. Solve for X.

17G.) $X^2 - 2X + 3 = 6X^2 - 2X + 5$.

17H.) $16^X = 9871$: Solve for X.

17I.) $24X/X^{-1} = 72$ Solve for X.

17J.) A circle has a (3,7) location on its circumference and it has a radius that is 2.5 meters long. Does this circle have any negative X values of its circumference and if so, how many?

NOTES

Part Eighteen – Impossible Mathematics

Example One

   John is twenty-eight years old and his wife, Juanita, is one year older. Their daughter is one fourth the age of John. When will her age be one quarter of the combined ages of her two parents? The answer, if I got it correct, is fourteen and one-half years.

   Let D represent the daughter, Jn represent John, and Ja represent Juanita. Since everybody gains a new year, even those who have stopped living, we can measure that yearly increase with the letter X. Also, let D be in the numerator and the sum of Jn and Ja be in the denominator of the fraction. On the right side of the equation place the desired one quarter.

$$\frac{D + X}{(Jn + X + Ja + X)} \quad = \quad 0.25$$

$$\frac{7 + X}{(28 + X + 29 + X)} \quad = \quad 0.25$$

$$7 + X \qquad = \quad 0.25(28 + X + 29 + X)$$

$$7 + X \qquad = \quad 7 + X/4 + 7.25 + X/4$$

$$7 + X - 0.5X \quad = \quad 14.25$$
$$0.5X \qquad = \quad 7.25$$
$$X \qquad = \quad 14.50 \text{ years or } 14.5 \text{ years.}$$

Back to the question:

The new age for the daughter is 7(the starting point of the daughter's years) + 14.5 years = 21.5 for the numerator. The new ages of the parents are 28 + 14.5 or 42.5 for the father and 29 + 14.5 or 43.5 for the mother, which will be combined in the denominator of the fraction. $\dfrac{21.5}{86}$ = 0.25 **Here, the daughter's age is one quarter of the sum of the parent's combined age.**

Example Two

This technique does not always work when one wants to solve a similar problem. For example, when the ages of two lives are compared to an age of a third life and we want to find out when will the third life be one half the total years of the other two lives there is a problem with the above techniques with solving this type of problem.

An example of this is where John is Fourteen years old and his father's gift automobile to him is twenty years old. John's little brother Sam is six years old. When will Sam be half the combined age of some future combined ages of John and his gift of an automobile that came from John's father?

$$\frac{6 + X}{14 + X + 20 + X} = 0.50$$

$$\frac{6 + X}{34 + 2X} = 0.50$$

$$6 + X = (0.5)(34 + 2X)$$
$$6 + X = 17 + X$$
$$X = 11 + X$$

$$X - X = 11$$
$$0 = 11$$

The problem is that we cannot solve for X by subtracting X from each side of the equation. In fact, if we add one million years to everybody's age we will see the following;

$$\frac{1{,}000{,}006}{1{,}000{,}014 + 1{,}000{,}020} = 0.50$$

$$1{,}000{,}006 = (0.5)(2{,}000{,}034)$$

Here the value on the left side of the equal sign never reaches to be one-half the value that is on the right hand side of the equal sign.

$$1{,}000{,}006 = (1{,}000{,}017)$$

Example Three

What happens if we try numbers that are above and below one-half, 0.50.

Now let's allow Mike to be six years old. His older brother is eight years old and his father's house is ten years old. When will Mike be fifty-one percent of the sum of his older brother, Phillip, and his father's house ages?

$$\frac{6 + X}{(8 + 10 + 2X)} = 0.51$$

$$6 + X = 0.51(8 + 10 + 2X)$$
$$6 + X = 9.18 + 1.02X$$
$$-3.18 = 0.02X$$
$$X = -159$$

Here, $\frac{6 + X}{(8 + 10 + 2X)} = 0.51$ and

$-153/-300 = 0.51$ So, 159 years ago Mike's age was fifty-one, 51, percent of the sum of his older brother's age combined with his father's house that was ten years old when he, Mike, was six years old.

Example Four

Now let's allow Mike to be six years old. His older brother is eight years old and his father's house is ten years old. When will Mike be forty-nine percent of the sum of his older brother, Phillip, and his father's house ages?

$$\frac{6 \quad + \quad X}{(8 + 10 + 2X)} = 0.49$$

$$6 \quad + \quad X \quad = (0.49)(18 + 2X)$$

$$6 \quad + \quad X \quad = 8.82 + 0.98X$$

$$0.02\,X \quad = \quad 2.82$$

$$X \quad = \quad 141$$

$$\frac{6 \quad + \quad 141}{(18 + 282)} = \frac{147}{300} = 0.49$$

Going further, when the number on the right hand of the equation, such as 0.50 or fifty percent, is the inverse of the number of members of the group, such as two members, the solution cannot be found, when the number on the right hand of the equation, such as 0.3333 or thirty-three percent, is the inverse of the number of members of the group, such as three members, the solution cannot be found, and when the number on the right hand of the equation, such as 0.25 or twenty-five percent, is the inverse of the number of members of the group, such as four members, the solution cannot be found. An exception to this is that when the number on the right hand of the equation, such as 1.00 or one hundred percent, is the inverse of the number of people in the group, such as there being only 1.00 member, the solution can be found. In the above situation, we have seen that when the percentage is above fifty percent, such as fifty-one percent, or below fifty percent, such as forty-nine percent, the solution is possible. Yet, when the percentage is fifty percent, and the number of members of the group that is represented in the denominator is two or the inverse of fifty percent, there is no solution.

To help me solve this pattern I went to see Assistant Professor Brenton A. Webber of Community College of Philadelphia, Pennsylvania. He told me that some equations show inconsistencies and, here, the pattern noted in example two will occur but it will not occur when there is not an inconsistency. Moreover, most, or many, students who graduated from college, after taking mathematics courses do not know of these facts.

A pattern is noticeable. It has been found that if $(cX_N)(Y)^{-1} = cX_d$

 Here c equals coefficient of X, d equals the denominator, and n equals the numerator. Y is the percentage, such as fifty percent in Example Two, which is wanted in the relationship. Under these conditions the unknown element, X, cannot be found.

For example if there are four elements in the numerator and sixteen elements in the denominator the solution for X cannot be found.

$$\frac{6 \quad + \quad 8 \quad + \quad 10 \quad + \quad 4 \quad + \quad 4X}{1 + 2 + 3 + 4 + 5 + 6 + 7 + 8 + 9 + 10 + 11 + 12 + 13 + 14 + 15 + 16 + 16X} = 0.25$$

$$\frac{28 \quad + \quad 4X}{136 + 16X} = 0.25$$

$$28 \;+\; 4X = 0.25(136 + 16X)$$

$$28 \;+\; 4X = 34 + 4X$$

$$4X - 4X = 34 - 28$$

$$0 = 6$$

Part Eighteen Problems

Do the following problems without use of a calculator

1.)   Marisa is four times her brother's age and half her mother's age. The sum of her parent's ages plus her and her brother's ages will sum to be one hundred five years from now. When will her and her brother's ages be, or when was her and her brother's ages, one fourth of the sum of her parents' ages?

2.)   In the same family of the above question, when will, or was, her age twenty-nine and three-fifths of the sum of her parents' ages?

3.)   When will, or was, her younger brother's age one half of her age?

4.)   Divide 123.2009 by -6.0040482 and place four decimal places in your answer after rounding it to the ten-thousandth decimal place.

5.) Rodney is Seventy-Eight Years old, today.   His mother is one hundred years old today and his father is one hundred and two years old. Only his mother is still living. When will, or was, Rodney's age be one half the sum of his parents' ages?

6.)   Amazu, which means that no one knows everything, has a father who is forty-nine years old. He is four years old and his sister is two years old. When, in the future, will he be one fifth of the sum of his father and his sister, at that time, future ages?

7.)   When will Amazu's age be one-half the sum of his father and his sister ages?

8.)   Looking at the above question, when will Amazu's father be five times his children's combined ages?

9.)   The value of a barn tractor depreciates at a rate of 1.034 a year. When will it be half the value that it had just after it was produced twelve years ago?

10.)   A circle has a (37, 20) location on it circumference. What must be the value of this circle's diameter?

Part Nineteen
Various Issues

The Stop - Start Function

St = the period when the function Starts
Sp = the period when the function Stops
N = the period at the present time

$$\left(\frac{\dfrac{2^{(St-n)}/2^{(St-n-1)}}{2^{(St-n)^2/2}/2^{(St-n-1)^2/2}}}{3}-1\right)\left(\frac{\dfrac{2^{(St-n)}/2^{(St-n-1)}}{2^{(St-n)^2/2}/2^{(St-n-1)^2/2}}}{3}-1\right)$$

Reducing the terms of the numerator gives

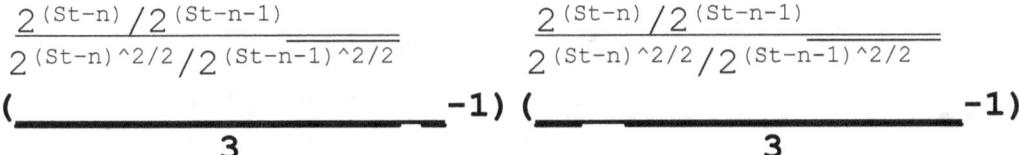

$$\left(\frac{\dfrac{2}{2^{(St-n)^2/2}/2^{(St-n-1)^2/2}}}{3}-1\right)\left(\frac{\dfrac{2}{2^{(n+1-sp)^2/2}/2^{(n-sp)^2/2}}}{3}-1\right).$$

Let's use the seven times table by multiplying it by each month's position, n, where January's position equals one, February's position equals two, March's position equals three, etc. Also, let this relationship start at the fourth month of the year and stops at the sixth month of the year. Last, sum up the values of f(x) times the Stop - Start Function.

Here, we will look at March, April, May, June, and July. March is being observed because it is a month that is outside this period of activity. Also, July is outside this period of activity. Your job is to replace the terms of the formula with the correct month's number.

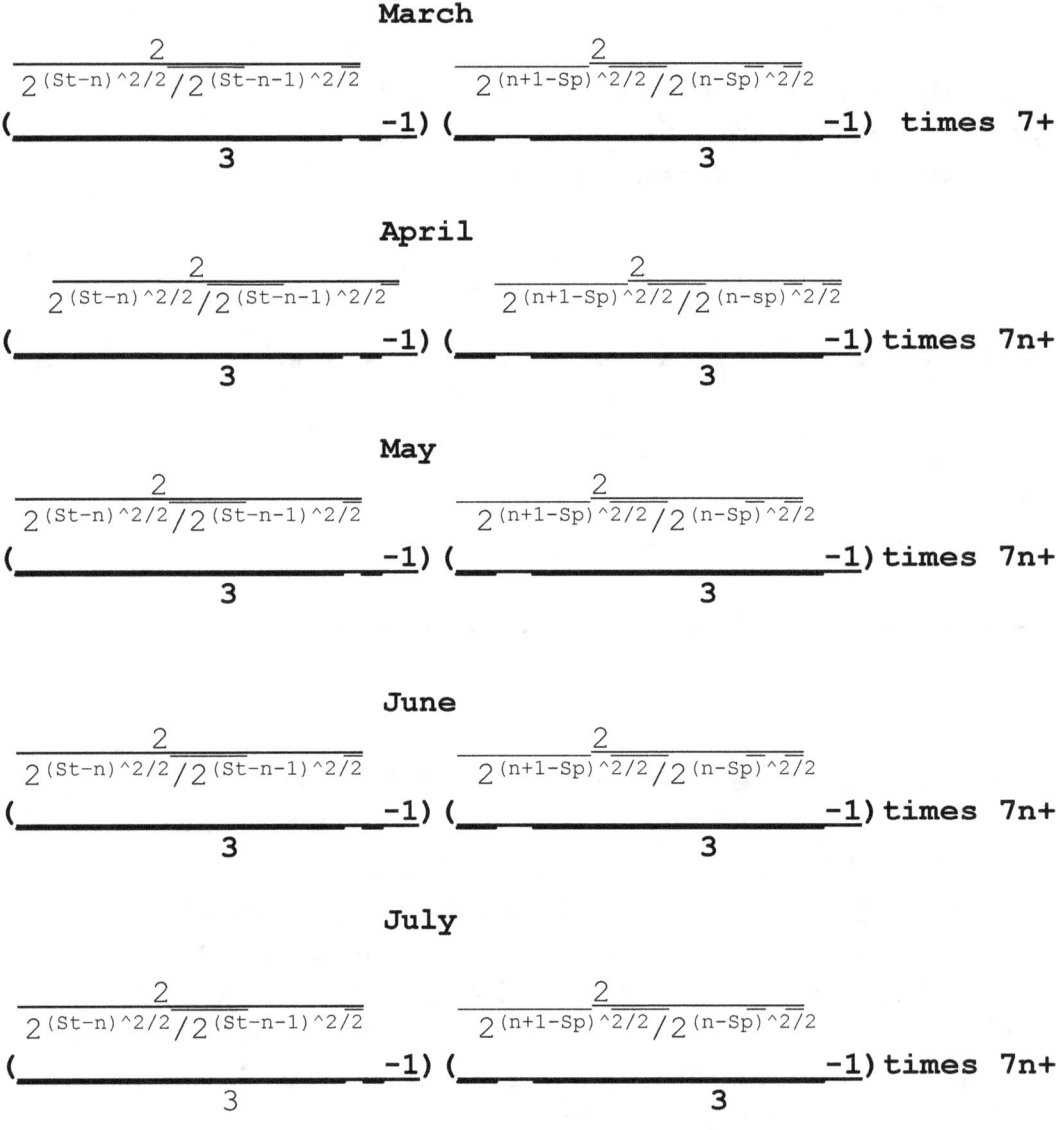

March

$$\left(\frac{2^{(St-n)^2/2}/2^{(St-n-1)^2/2}}{3}-1\right)\left(\frac{2^{(n+1-Sp)^2/2}/2^{(n-Sp)^2/2}}{3}-1\right) \text{ times } 7+$$

April

$$\left(\frac{2^{(St-n)^2/2}/2^{(St-n-1)^2/2}}{3}-1\right)\left(\frac{2^{(n+1-Sp)^2/2}/2^{(n-sp)^2/2}}{3}-1\right) \text{ times } 7n+$$

May

$$\left(\frac{2^{(St-n)^2/2}/2^{(St-n-1)^2/2}}{3}-1\right)\left(\frac{2^{(n+1-Sp)^2/2}/2^{(n-Sp)^2/2}}{3}-1\right) \text{ times } 7n+$$

June

$$\left(\frac{2^{(St-n)^2/2}/2^{(St-n-1)^2/2}}{3}-1\right)\left(\frac{2^{(n+1-Sp)^2/2}/2^{(n-Sp)^2/2}}{3}-1\right) \text{ times } 7n+$$

July

$$\left(\frac{2^{(St-n)^2/2}/2^{(St-n-1)^2/2}}{3}-1\right)\left(\frac{2^{(n+1-Sp)^2/2}/2^{(n-Sp)^2/2}}{3}-1\right) \text{ times } 7n+$$

$$= 0 + 28 + 35 + 42 + 0 = 105$$

Make sure your numbers agree with the final answer. The left side of this formula handles periods before the start of the function and the right side of the formula handles periods after the stop of the function.

179

<u>Notes</u>

Dance of the Tree

Six moles of Carbon Dioxide

Six moles of Water

Makes one Mole of Glucose & Six of a diatomic molecule named Oxygen

    This Process is Photosynthesis

    Parenchyma, Cholorenchyma, and Sclerenchyma

    Cells of a Tree

        At Twenty-First and Tioga.....Streets

        Who hides behind Leaves, Branches, & a Truck

Like Human Beings

They show changes in Time

We We We

We can We can * Run

& try ta out race the Tree

But it gets here first with Morning Due

    We can spin & spin in rotations

    But its Dance determines the spin of seasons & positions of planets

        While it Changes Colors

        Changes Colors of Leaves

& & & &

& & & & when Showers

Sky Showers Us

& when warm Sun warms Us

We hide under its Beautiful

Shades

Lights of Life

    We donate Carbon Dioxide

    It gives back with Oxygen

    Sky donates Water

    It returns with Glucose

    Fruit, Fiber, Fresh Air

    All members of Life

Trees at Twenty-First and Tioga   Streets

Lights of Life

Like those all around the World

Are here

We must never take You for granted

& We will never forget You

Take Us to Your Tomorrows   Kesho

SUMs SUMs

Should we list numbers, that differ by a constant amount, of a multiplication table in a column and place a copy of the first number of that series of numbers in a second column before adding it to the second number of the first column and placing the result of that into the second vertical line's next position and continue to do this process for a number of times before creating additional vertical columns in the same manner as what was done with the second column, the results we would created would become a chart of vertical and horizontal lines. How can we find the value of any horizontal and vertical place of this chart? We can find it by using the Sum's Sums Formula.

We can have a chart that looks like the below chart when the chart's width is three, 3, and when the chart's length is four, 4. The formula on the right of the chart is used to find particular numbers of the chart and numbers of particular locations of the chart.

```
                    W I D T H

                  1       2       3

  L  1       8.12   8.12    8.12    8.12
  E
  N  2       16.24  24.36   32.48   40.60
  G
  T  3       24.36  48.72   81.20   121.80
  H
     4       32.48  81.20   162.40  (284.20)      (L+W)!
                                             (K)-------------------
     5       40.60  121.80  284.20  568.40       (L-1)!(W+1)!
```

$$\frac{(8.12)(3+4)!}{(3)!(4)!} = \frac{(8.12)(7)(6)(5)(4)(3)(2)}{(3)(2)(4)(3)(2)} = 284.20$$

	0	1	2	3	4
01	6.35	6.35	6.35	6.35	6.35
02	12.70	19.05	25.40	31.75	38.10
03	19.05	38.10	63.50	95.25	133.35
04	25.40	63.50	127.00	222.25	355.60
05	31.75	95.25	222.25	444.50	800.10
06	38.10	133.35	355.60	800.10	1,600.20
07	44.45	177.80	533.40	1,333.50	2,933.70
08	50.80	228.60	762.00	2,095.50	5,029.20
09	57.15	285.75	1,047.75	3,143.25	8,172.45
10	63.50	349.25	1,397.00	4,540.25	12,712.70

This chart has ten rows and columns numbered zero to four.

To find the value of any cell one can use the following formula: $\dfrac{K(L + W)!}{(L-1)!(W+1)!}$

Here, the letter **K** stands for the value of the multiplication table. The letter **L** stands for the length of this table that is used. The letter **W** represents the width of the table that is used.

Should you divide all the numbers of this chart by **K**, or 6.35, and make a "new column that only has the number one in it, you will see that the chart's pattern is really Pascal's numbers and has the Pascal's numbers on a slant that "runs" from the column with the number of ones up the page and on a slant.

	New Column	0 column	1 column	2 column	3 column	4 column
Row 1	1	1	1	1	1	1
Row 2	1	2	3	4	5	6
Row 3	1	3	6	10	15	21
Row 4	1	4	10	20	35	56
Row 5	1	5	15	35	70	126
Row 6	1	6	21	56	126	252
Row 7	1	7	28	84	210	462
Row 8	1	8	36	120	330	792
Row 9	1	9	45	165	495	1,287
Row 10	1	10	55	220	715	2,002

ACCELERATION

(A) If the first interest payment of (P&I), the monthly payment of a loan or mortgage that does not include taxes and insurance, is subtracted the remainder will be P_1. It is

A payment of $0 is made on a loan, or mortgage.

This amount, P_0, will be subtracted from the balance of the mortgage, and interest cost will be applied to this new balance.

$$\text{(P\&I)} - (P^* - P_0)(\text{Int}/12) = P_1 \text{ ,}$$

where P_0 is a payment of zero dollars.

This is

A.) **The first principal payment on a loan, or mortgage.**

This amount, P_1, will be subtracted from the balance of the mortgage, and interest cost will be applied to this new balance.

$$\text{(P\&I)} - (P^* - P_0 - P_1)(\text{Int}/12) \quad = P_2,$$

Where P_1 is the first payment against the principle value of the mortgage.

This is

B.) **The second principal payment on a loan, or mortgage.**

This amount, P_2, will be subtracted from the balance of the mortgage, and interest cost will be applied to this new balance.

$$\text{(P\&I)} - (P^* - P_0 - P_1 - P_2)(\text{Int}/12) \quad = P_3 \quad \text{where } P_2 \text{ is this second payment.}$$

This is

c.) The third principal payment of a loan, or mortgage.

This amount, P_3, will be subtracted from the balance of the mortgage, and interest will be applied to this new balance.

$$(P\&I)-(P^*-P_0-P_1-P_2-P_3)(Int/12)= P_4.$$

The left sides of the equations,　　　　A.) to C.), equal each other.

Thus, the right side of each equation,　A.) to C.), equal each other.

After equating equations　　　　　　　A.) to C.), we can see that

To go from P_2 to P_3, for example, we must include a $(-P_2)(Int/12)$ term. This term is found by subtracting P_2 from P_3. Also, a $(-P_2)$ term must be included to show the difference between there two terms because we do not just want the interest value of $(-P_2)(Int/12)$, but we want the principal value, also. This principal value can be factored out of the parentheses. This gives the result of

$P_3 = P_2 (1+ Int/12)$

or　$P_2 = P_1 (1+ Int/12)$

or　$P_3 = P_1 (1+Int/12)(1+ Int/12)$

Of course, a loan or mortgage has, often, many payments, such as P_1 to P_n. The sum of individual payments equals the amount of funds that was borrowed, starting with P_1 and ending at P_n.

Again,

$P_1(1+Int/12)^0 + P_1(1+Int/12)^1 \ P_1(1+Int/12)^2+ \ldots \ldots P_1(1+Int/12)^n$

$= P^*$

For a small time, let us allow $(1+Int/12)$ to equal $[A]$.

Here,

$([[A]]^{(n+1)} -1)/(A-1) = P^*$ or the borrowed amount.

This technique can also be used to find the value of any continuous relationship of a number and its exponents. Let us look at the number three.

$$3^0 + 3^1 + 3^2 + 3^3 + 3^4 = 3^{(4+1)}-1/(3-1) = 121$$

$$1 + 3 + 9 + 27 + 81 = 121$$

Thus, the answer for the sum P_1 can be found, easily.
Here, we can increase the amount of P_1 to a value greater than the minimum amount that the bank or lender dictated to us and see that thousands of dollars, of many loans and mortgages, can be easily saved, or spent on other items other than bank interest payments and bank profits.

$ 100,000.00
$ 100,000.00

Mortgage Acceleration

The principle can be extended from that of a loan or mortgage to that of an investment. Here, the same mechanism that allows you to save on a mortgage is the principle that can allow you to invest in some other monetary relationship, such as mutual funds.

Acceleration Example

Here, let the goal, that is the amount of money that you want to save, be "P_1." This symbol was used before. Now, you are in charge of obtaining the goal, your goal. In this example, the goal is to save, about, $100,000 plus inflation, after making monthly deposits for twenty years.

Here, there are 240 monthly investments, the earned interest rate is twenty percent, the inflation rate is four percent, and the monthly investment amount is sixty dollars.

$$\frac{P_1\left([1+Int-Inf/12]^{241} - 1\right) = P^*}{(Int-Inf)/12}$$

Here (A) is (1+Int-Inf/12)

$$\frac{\$60\left([1+(0.2-0.04)/12]^{241} -1\right) = P^*}{(0.2-0.04)/12}$$

Again, inflation is working at four percent and mutual funds, or some other investment, earn twenty percent interest, yearly. Thus, your earned interest will be twenty percent minus the four percent inflation rate. Inflation is symbolized by the term Inf. Thus, the interest that you earn = Int-Inf, (0.20-0.04). = 0.16

$$\frac{P_1\left([1+(Int-Inf)/12]^{N+1} - 1\right) = P^*}{(Int-Inf)/12}$$

$$\frac{(\$60)\left[(1+[0.16]/12)^{241} -1\right]}{(0.16)/12} = \$105,027.65$$

T h e R u l e O f 72

Many investments grow at a rate where they double in value every seventy-two periods, or months. This equals a percentage growth rate of about 12.5%. Here, divide twelve and one-half percent by the twelve months of a year and then add one to the result before enclosing the results by parentheses and then raising the results to the seventy-second, 72, power, $(1 + 0.125/12)^{72}$. What is the purpose of the one that is added to 0.125/12?

One allows the sum, that includes it and that is greater than one, to be increased when it is raised to a power that is greater than a power of one.

Questions 57-60

Ms. Bart was approved to receive a fifty-two thousand two hundred dollars mortgage. The interest rate on the mortgage is seven percent and she has twenty-four years to pay the mortgage. Her house will have a monthly cost that will include taxes and other costs that are two thousand seven hundred and eighty-six dollars and twenty-eight cents plus the amortization of the fifty-two thousand two hundred dollars financing of the cost of the house that she purchased after paying a ten percent down payment on the sale of the house.

57.) What was the sale price of the house before she made any down payment?

58.) What will be her monthly amortized mortgage payments?

59.) In order to pay this mortgage within twelve years how much extra money must she pay monthly?

60.) Triangle D has Point A which is at (-16, 8), Point B which is at (21, -1), and Point C which is at (13, 0). Across from Point A is side a, across from Point B is side b, and across from Point C is side c. Find the degrees of these three points of Triangle C.

_____

_____

Part Nineteen Problems

19A.) What is the value of the fifth row and fourth column of a Sum's Sums chart that has five rows and six columns where the difference between each item of the zero column is 3.88?

19B.) Amortize a car that comes with a fifty month financing arrangement of a twenty-six thousand dollar loan with an interest cost of six and one-half percent.

19C.) Amortize a boat that comes with a seventy-two month arrangement of fifty-one thousand five hundred dollars to be financed at a three percent interest over ninety payment months.

19D.) How much extra must the person/s in question 198 pay each period to finished the note at the thirty-eighth payment period after the payer of this agreement paid, on time, four monthly payments?

190E.) A circle has a diameter of 1447.5 centimeters and an X value of 3 meters. Here, this circles Y value is what?

Questions

61.)   Matrix Multiplication

$$
\begin{bmatrix}
2 & 0 & 0 & 9 \\
-2 & -4 & 10 & 11 \\
15 & -2 & 8 & 22 \\
0 & -3 & 6 & 99
\end{bmatrix}
\text{ times }
\begin{bmatrix}
-2 & 22 & 9 & -6 \\
0 & 22 & 9 & 3 \\
-9 & 4 & 0 & -2 \\
0 & 3 & 2 & 1
\end{bmatrix}
$$

In 2004, Dr. Mark Dean was selected as one of the 50 Most Important Blacks in Research Science

Dean has been with IBM since 1980. Dean holds 3 of the original 9 patents on the computer that all PCs are based upon: Soon after joining IBM, Dean and a colleague, Dennis Moeller, developed the interior architecture (ISA systems bus) that enables multiple devices, like modem and printer, to be connected to personal computers. Then he worked for a number of years before considering the doctorate.

99999999999999999999999999999944444444444444444444444444444488888888

Part Twenty (Answers to Questions)

When the quadratic formula gives no solution you can graph the function. There you will see that the function never reaches a value of where y = zero.

To find a simple **derivative** many functions one only needs to multiply each element by the elements exponent and replace the old exponent with a new exponent that is one less than the old exponent. For example $4Y^6 - 2X^4 - 9$ has a derivative value of the $24Y - 8X - 0$ because $4Y^6$ becomes $24Y^5$ once we multiply $4Y$ by the exponent of six, $-2X^4$ becomes $-8X$ because the -2 of this element was multiplied by the exponent of 4 before that exponent was replaced by the number three, which is one less than the original exponent of four. the integers, including the number nine in this problem, have no X value/s so when we get the derivative, or change in X, of it we have to arrive at zero because its movement of X, here, is non-existent because it had, and has, no variable X. With functions like $2X^2 + 3 = Y$ we will see a similar relationship to what is in the following chart.

$2X^2 + 3 = Y$

X	Y Values	The differences between neighboring Y values	The differences between the differences of neighboring Y values	
-1	5	-2	4	
0	3	2	4	
1	5	6	4	
2	11	10	4	
3	21	14	4	
4	35	18	4	
5	53	22	4	

The derivative of the function is 4X and if you get the derivative of that you will arrive at the fourth column's constant value of four, 4.

195

01.) On a graph draw the function $Y = X^2 - X - 2$. See the graph.

One will see that the derivative of this function is $2X - 1 = 0$ after it is set to equal zero. This technique is used to show a function's extreme. Here we can see that the function has an extreme where one and one-half equals X and the resulting Y value is two and quarter.   Here is the graph of $X^2 - X - 2 = Y$.

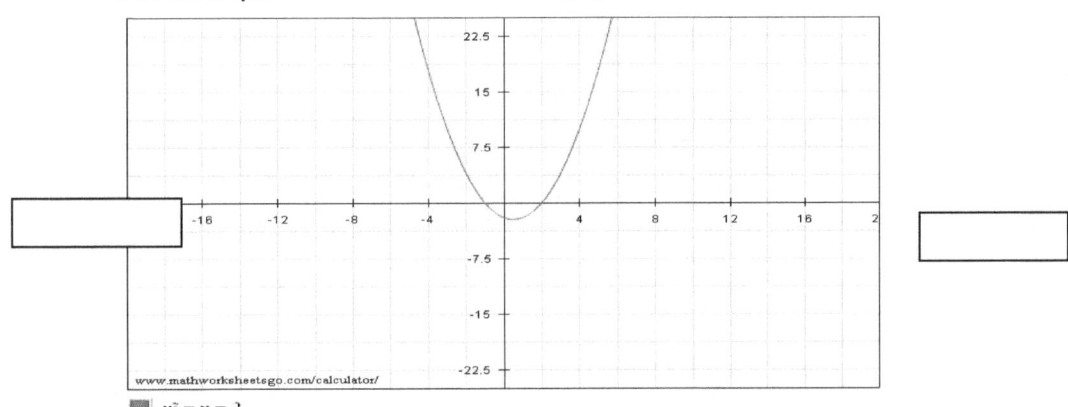

www.mathworksheetsgo.com/calculator/

$x^2 - x - 2$

Y' = 2X -1 is

the derivative function of this expression. Solving for when this function is zero we have the following results.

$0 = 2X - 1$, $2X = 1$, So $X = 0.5$ and $(0.5)^2 - 0.5 - 2 = -2.25$. So, at $X = 0.5$ and Y equals -2.25 the function is at the lowest point.

02.) -66.08 + 43.23 -4.09 + 1,008.19 = 981.25

03.) 34-54 = 66 + X What is the value of X?   Subtract   66
from both sides of the equation.   34 - 54 -66 = 66 – 66 + X
-86 = X

04.) 456.98 is what percent of 200,200?
456.98/ 200,200 = 0.00228261738 so 456.98 is 0.228261738 percent of 200,200.

05.) Solve for X when the equation is $3X^2 - 2X + 12 = 16$ See the Graph

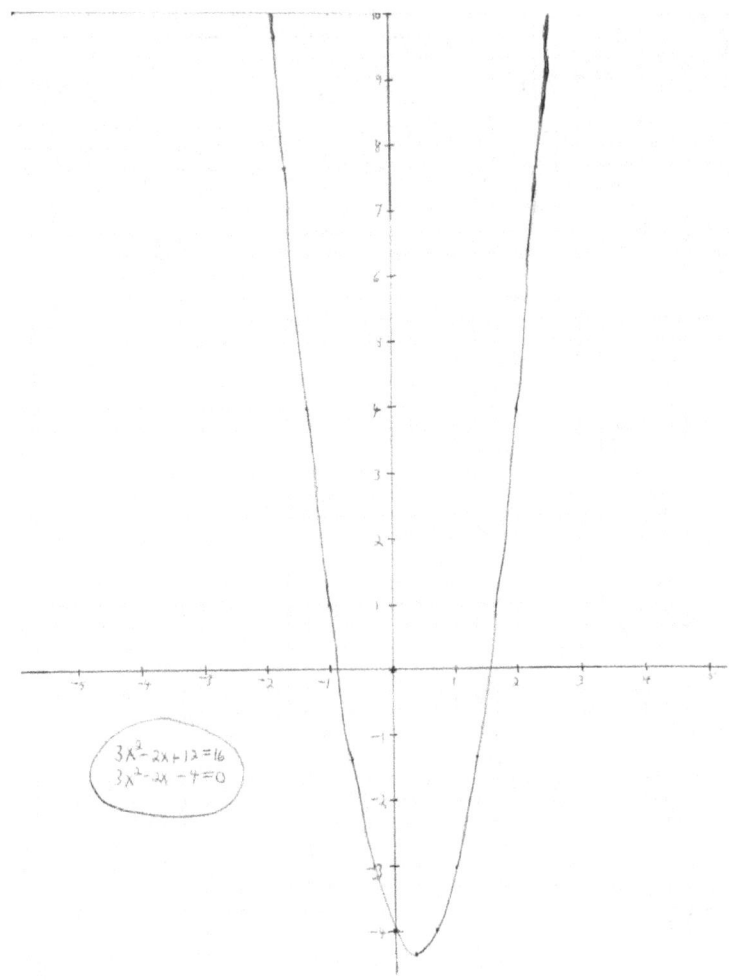

3X² – 2X + 12 = 16 is the same as 3X² – 2X – 4 = 0.   Use the binomial formula, -b +/-(b²-4ac)$^{0.5}$/2a, to graph this function.

$$[2+/- ([-2]^2 + 48)^{0.5}]/6 \ = \ [2 + 7.211102551]/6 \ \ = \ \ 1.535183759$$
$$= \ [2 - 7.211102551]/6 \ \ = \ -0.868517091$$

So, (X - 1.535183758)(X + 0.8685170912 = 0

X² – 0.66666666666X – 1.33333333333 = 0   Multiplying this by three gives us back our original equation of 3X² – 2X – 4 = 0 which is also, after adding 16, 3X² – 2X + 12 = 16.

197

Another method is to get to where the equation becomes $3X^2 - 2X - 4 = 0$ and divide this by the coefficient of X^2. From there it will be easy to use the quadratic formula.

06.) Change 3/57 into a percent and into a decimal number.

3/57 = 1/19 after dividing each part of the fraction by three. 1/19 = 0.052631. So as a percent the answer is 5.2631 percent and, again, as a decimal number the answer is 0.052631.

07.) Change 0.0123 into a percent and into a fraction.

It can be seen as being 1.23 percent after we move the decimal point two places to the right. It also can be seen as 123 ten thousandths, 123/10,000, because the fourth decimal space is the ten thousandths space.

08.) 92/46 − 622/34 =   (68 -622)/34 = -554/34 = -16.2941

This was done by seeing that the first fraction is really two. Then, it was decided to give two a denominator of 34. Since the denominator of the number two, which is the number one, goes thirty-four times into thirty-four and that result times the numerator of two, which is the number two, equals sixty-eight we know that progress has been made with solving this problem. So the first fraction became 68/34 and the second fraction remained -622/34. So, both of these fractions have a common denominator and they can be viewed as the following problem.

09.) (68 − 622)/34 and this equal -554/34 which also equals -16.2941.

10.) $456(5.05)^2 - X = 25.0500$ and solve for X.

(456)(25.5025) − X = 25.05   11,629.14 - 25.05 = X

11,604.09   = X

11.) 5/56 + 567/50 + 34/56 + 10/45 =

We can combine, easily, the numbers that have 56 as the denominator to reach 39/56. As a decimal the remaining fractions of 39/56 + 567/50 + 10/45 are

0.696428571

  11.340000000

+ 0.222222222

  12.25865079 or 12.2587

12.) -45/78 + 23X/39 + 452/7 -45/78 = 0, solve for X.

We can give the second fraction the denominator of 78 and multiply it's numerator by the number of times 39 goes into 78. Here, the problem becomes

  -90/78   + 46X/78   + 452/7   = 0

  -1.153846154 + 0.589743589X   + 64.57142857   = 0

  63.41758242 + 0.589743589X   = 0

Next, we subtract the first number of the above equation from both sides of the equation and the result is

0.589743589X   = 63.41758242

After dividing both sides of the last equation by the first number in the equation the results are   X = 63.41758242/0.589743589   = 107.5342078

13.) (-56/78)(78/1) =

Here we only need to multiply the numerators and the denominators before dividing the new denominator into the new numerator. We also can see that the numerator of 78 cancels the denominator of 78 until they become a number one in the numerator and a number one in the denominator. Here, the only number that does not have a magnitude of one is the number -56, the answer.

14.) Answer with 8 decimal places (234)(4/61)^3 =

(234)(64/226,981) = 0.065979091 or 6.5979091^{-2}. Placing the first digit of a decimal number that has no value that is above zero on the left side of the decimal point can be balanced by raising the number to a negative power that represents how many times the decimal point has to be moved to create the answer.

A Newark, New Jersey Building

15.) $(2/21)(23/51) =$

       2 times 23 in the numerator and 21 times 51 in the denominator. The numerator becomes 46 and the denominator becomes 1,071. Last the numerator divided by the denominator shows the answer to be 0.042950514 or 0.04295

16.) $(6/201)(33/441)$ $=$ $198/88,641$ $= 0.002233729$

       Also, by moving the decimal point in a negative, or right, direction for three steps we see the answer is 2.233729^{-03}. When expressing answers, with this technique, the answer should have one digit on the left hand side of the decimal point.

17.) $X^2 - 12X - 15.75$ Solve for X.

       Use the quadratic formula $-b +/- (b^2 - 4ac)/2a = 0$ where a, b, and c are the first, second and third coefficients of a normally set equation that has the format of $aX^2 + or - bX + or - C$ Here we will have this equation. $12 +/- (144 + 63)/2$ The results of this gives us two answers, $(13.1937, -1.1937)$.   The value of X has both of these options, in order for the sum of X with the negative of either one of these two options to be zero. So we have $(X - 13.1937)(X + 1.1937) = 0$.

       Using the FOIL method of multiplication will give us the original equation, again. $(X - 13.1937)(X + 1.1937) = X^2 - 12X - 15.75$.

18.) $2X(2-X^2) + 2X/(X)^3 = (2X)(2 - X^2) + (2X)(X^{-3})$ Solve for X.

       This "boils down" to be

       $2X(2-X^2) + (2X)(X^{-3}) = (2X)(2 - X^2) + (2X)(X^{-3})$

We can subtract $(2X)(X^{-3})$ from both sides of the equation and the results are written below.

       Since $2X(2-X^2)$ is the remainder on both sides of the equation, the equation cannot be solved.

19.) $20/65 - 40/130$ $=$ $40 - 40/130$ $= 0/130$ $= 0$

20.) $37X^4$ $89 - 25X/(2X)^{-3} = 52,688$

      First, $3,293(X^4) - 25X(2X^3) = 52,688$ then $(X^4)(3,293 - 25) = 52,688$ then $X^4(3,268) = 52,688$ then $X^4 = 52,688/3,268$ then X^4 $= 16.12239902$ then $X = (16.12239902)^{0.25}$ and then $X = 2.003814045$ or 2.003814.

21.) Solve for X when the equation is $4X^2 - 20X + 5 = 10.2$

$4X^2 - 20X - 5.2 = 0$ is a result from subtracting 10.2 from both sides of the equation.
The Quadratic Equation will be used to solve for X after the equation is divided by four and becomes $X^2 - 5X - 1.3 = 0$.

$-b +/- (b^2 - 4ab)/2$ or $[5 +/- (25 + 5.2)^{0.5}]/2 = [5 +/- (30.2)^{0.5}]/2$

$[5 +/- 5.495452666]/2$

This is the same as 5.247726333 or -0.247726332.   To
check this make X have these values.   (X –
5.247726330)(X + 0.247726332)   This equals $X^2 - 5X + 1.3 = 0$ and when this entire equation is multiplied by 4 the relationship of the two sides of the equation remain the same as the equation becomes what we had in the beginning, $4X^2 - 20X + 5 = 10.2$. We now see that the value of X is 5.247726333 and it is also near - 0.247726332.

22.) Change 3.03 in to a percent.

By moving the decimal point two places to the right we arrive at 303 percent, 303%.

23.) Solve for X when the equation is $2X^2 - 4X + 4 = 2$.

Here, we can see that $2X^2 - 4X + 4 = 2$...becomes ...$2X^2 - 4X + 2 = 0$...which becomes...$X^2 - 2X + 1 = 0$ can be solved with use of the quadratic formula,

$[2 +/- ([-2]^2 - 4(1)(1))]^{0.5}/2 = 0$. So, this gives us $(2 +/- 0)/2 = 0$. After dividing by two we can see that the only possible answer is one.

$(X - 1)(X - 1) = X^2 - 2X + 1$ Multiplying two to this result gives us the original equation of $2X^2 - 4X + 2. = 0$

24.) Solve for X when the equation is $4X^2 - 12X + 2 = -6$

Here, you can use the binomial formula or find two numbers that when multiplied together equal two, the last number of the equation, and when added equals negative three, -3, or the second integer of the equation.

$4X^2 - 12X + 8 = 0$ or $X^2 - 3X + 2 = 0$

$[3 +/- ([-3]^2 - 8)^{0.5}]/2$

$(X - 1)(X - 2)$ = 0 So, X can equal 1 and it can equal 2.

25.) 452.00 - 809.05 = -357.05

26.) $(2X - X^6)/X^{(8-2)}$ Simplify/reduce terms
          $(2X - X^6)/X^{(6)} = 2X/X^6 - X^6/X^6 = 2/X^5 - X^5/X^5 = (2 - X^5)/X^5$

What occurred here is that the denominator of X^6 is the denominator of 2X and it is the denominator of $-X^6$. While being the denominator of 2X is became X^5 because the numerator's X had an exponent of one. When dividing we are to subtract any exponents from like terms. What also occurred is that the other fraction that has a denominator of X^6 also has a numerator of X^6. So, we can name it anything that has a value of one, where anything divided by itself equals one. Thus it can be named X^5/X^5. Now we have two fractions that have a common denominator of X^5. The first one has a numerator of two, 2, and the second one has a numerator of X^5. So, simplified, this expression of

$(2X - X^6)/X^{(8-2)}$ becomes $(2 - X^5)/X^5$.

Do me a big favor, never forget this process.

27.) $5X - 2(X^2 - 4) = 4$, solve for X.
          $5X - 2X^2 + 8 = 4$
          $-2X^2 + 5X + 4 = 0$
          $X^2 - 2.5X - 2 = 0$
          $[2.5 +/- (-2.5^2 + 8)^{0.5}]/2 = 0$
          $[2.5 +/- (6.25 + 8)^{0.5}]/2 = 0$

The answer set that is X is (3.1375 and -0.6375) so to make an equation that will equal zero we have the below arrangement.
$(X - 3.1375)(X + 0.6375) = X^2 - 2.5 X - 2$. Note that the right hand side of this equation when multiplied by negative two equals the original equation, $-2X^2 + 5X + 4 = 0$.

28.) $22X [23/(4 + 16)] = 22$ Solve for X.

$$22X \; 23/20 \quad = 22$$
$$X \; 23/20 \quad = 22/22$$
$$X \; 23/20 \quad = 1$$
$$X \quad\quad\quad = 20/23$$

29.) $16X^2 - 9X + 4 = 3X$ Solve for X.

                 To solve for where these two functions intersect we can simply graph them. In the graph of each of them, $16X^2 - 9X + 4$ and $3X$, we can see that they never intersect, so there is no solution to the problem.

30.)   Change 0.0123 into a percent and into a fraction.

Moving the decimal point two spaces to the right gives 1.23 percent. Seeing four spaces tells us that we are measuring in one hundred thousandths. So, the fraction will be found by reading the numbers left of the zero as the numerator and the denominator as ten-thousandths. So it is one and twenty-three hundreds percent and it is 123/10,000 as a decimal number.

31.) $-2X - 44 \; = 22$ What is the value of X?

            To solve this add 44 to both sides of the equation

$$-2X - 44 + 44 \; = 22 + 44$$

$-2X \quad\quad\quad\quad = 66$ Then divide both sides of the equation by -2.

$$-2X/-2 \quad\quad = 66/-2$$

$$X \quad\quad\quad\quad = -33$$

32.) What is the distance between points (-25X, 204Y, 0Z) and (14X, 2Y, 13.98Z)?

$$[(-25-14)^2+(204-2)^2+(0-13.98)^2]^{0.5} = [(-39)^2+(202)^2+(-13.98)^2]^{0.5} =$$
$$[(1521)+(40804)+(195.4404)]^{0.5} = \quad 206.2048506 \text{ or } 206.2049$$

if we only use four decimal places in the answer.

33.) $(X^{0.3})^{20.2}X^2 =$ What   It equals $X^{8.06}$

For the order that operations should be done remember and follow the poem   .
  Please Excuse My Dear Aunt Sally.
  Here, P represents parentheses, E represents exponents, M represents multiplication,
  D represents division, A represents addition, and S represents subtraction.

  First, we multiplied the exponents of 0.3 and 20.2 together before we add the
  exponent that is two, 2. This was done because we are to work at the parentheses
  before we use any other exponent rule.

34.) <u>Kwanzaa Homework:</u>   <u>What is</u>   <u>Dr. Dean's relationship to the computer?</u>

Kwanzaa Homework

35.) -2X − 44 = 22   What is the value of X?

Add 44 to both sides of the equation

-2X − 44 + 44 = 22 + 44

-2X   = 66   Divide both sides of the equation by -2.

(-2/-2)X   = 66/-2   So, X   = -33

36.) Solve the following matrix

$$8A + 4B + 2C = -3$$

$$3A - 4B - C = 6$$

$$-3A - B + 2C = -3$$
9999999999999999999909999999999999999990009999999999

$$3A - 4B - C - 6 = 0$$

$$-3A - B + 2C + 3 = 0$$

$$8A + 4B + 2C + 3 = 0$$
8888888888888888888888888888888888888888888888888888

$$3A - 4 \quad B - \quad C - 6 = 0$$

$$-5 \quad B + \quad C - 3 = 0$$

$$14.6666B + 4.6666C + 19 = 0$$
3333333333333333333333333333333333333333333333333333

$$3A - 4 \quad B - \quad C - 6 \qquad = 0$$

$$-5 \quad B + \quad C - 3 \qquad = 0$$

$$7.6C + 10.2 \qquad = 0$$

C = -1.342105263

$$-5B - 1.342105263 - 3 = 0$$

$$-5B = 4.342105263$$

So B = -0.868421052

$$3A - 4(-0.868421052) - (-1.342105263) - 6 = 0$$

$$3A + 3.473684208 + 1.342105263 - 6 \qquad = 0$$

$$3A \qquad = 1.184210529$$

So A = 0.394736843

37.) $\dfrac{3B^3W^2 - 12B^6W^2}{2B}$ = when B = 1.4 and W = -2.0

$\dfrac{3(1.4)^3(-2)^2 - 12(1.4)^6(-2)^2}{2.8} =$ $\dfrac{3(2.744)(4) - 12(7.529536)(4)}{2.8}$

= $\dfrac{32.928 - 361.417728}{2.8}$ $\dfrac{}{2(1.4)}$

= $\dfrac{-328.489728}{2.8}$ = -117.31776

Another way of completing this problem can be done by noticing that the problem is

$\dfrac{3B^3W^2}{2B}$ - $\dfrac{12B^6W^2}{2B}$ because each term, or things that comes after a positive or negative sign, of the numerator is divided by the denominator.

38.) How many ways can we arrange the integers of 1382491?

There are seven factorial ways to arrange them. 7! = (7)(6)(5)(4)(3)(2)(1) = 5,040.

39.) What is the result of $5^{3!}$?

It is 5 to the sixth power or 15,625.

40.) -66.08 + 43.23 – 4.09 + 1,008.19 = 1,051.42 - 70.17 = 981.25

41.) 34-54 = 66 + X. What is the value of X?
To solve this place an additional -66 to both sides of the equation.
34-54 - 66= 66 + X - 66
34 -120 = X so -86 = X

42.) With four decimal places in your answer, if a circle has a diameter value of 8 feet what is the length of X when the length of Y is 3?

$$X^2 + Y^2 \quad = H^2$$

Since the circle's diameter is eight feet then the radius must be four feet. With this we know that Y has a value of three, 3. So, Y squared equals nine and the hypotenuse squared equals sixteen.

$$X^2 + 3^2 \quad = \quad 4^2$$
$$X^2 + 9 \quad = \quad 16$$
$$X^2 + 9 - 9 = \quad 16 - 9$$

$$X^2 \qquad = \quad 7 \qquad \text{So X equals} \qquad \underline{(7)^{0.5}}$$

or 2.6457.

43.) $-66.087 - 140 + 3.37 = \quad 202.717$

44.) $34 - 54 = 66 + X$. What is the value of X?

It is -86 and was found by subtracting sixty-six from the equation

45.) With four decimal places in your answer, if a circle has a diameter value of 18.1 meters what is the length of Y when the length of X is 0.5?

$$0.5^2 + Y^2 \qquad = \qquad 9.05^2$$
$$Y^2 \qquad = \qquad 9.05^2 - 0.5^2$$
$$Y \qquad = \qquad (9.05^2 - 0.5^2)^{0.5}$$
$$Y \qquad = \qquad (81.6525)^{0.5} \quad \text{So, Y is 9.036177289.}$$

46.) $\ln 64/\ln 16 = 4.158883083/2.772588722 = 1.5$

47.) Of the graphed triangles find the degree of each of the angles.
a.) Angle A=102.7521198 b.) Angle B=46.54815766 and
c.) Angle C= 30.69942623 degrees.

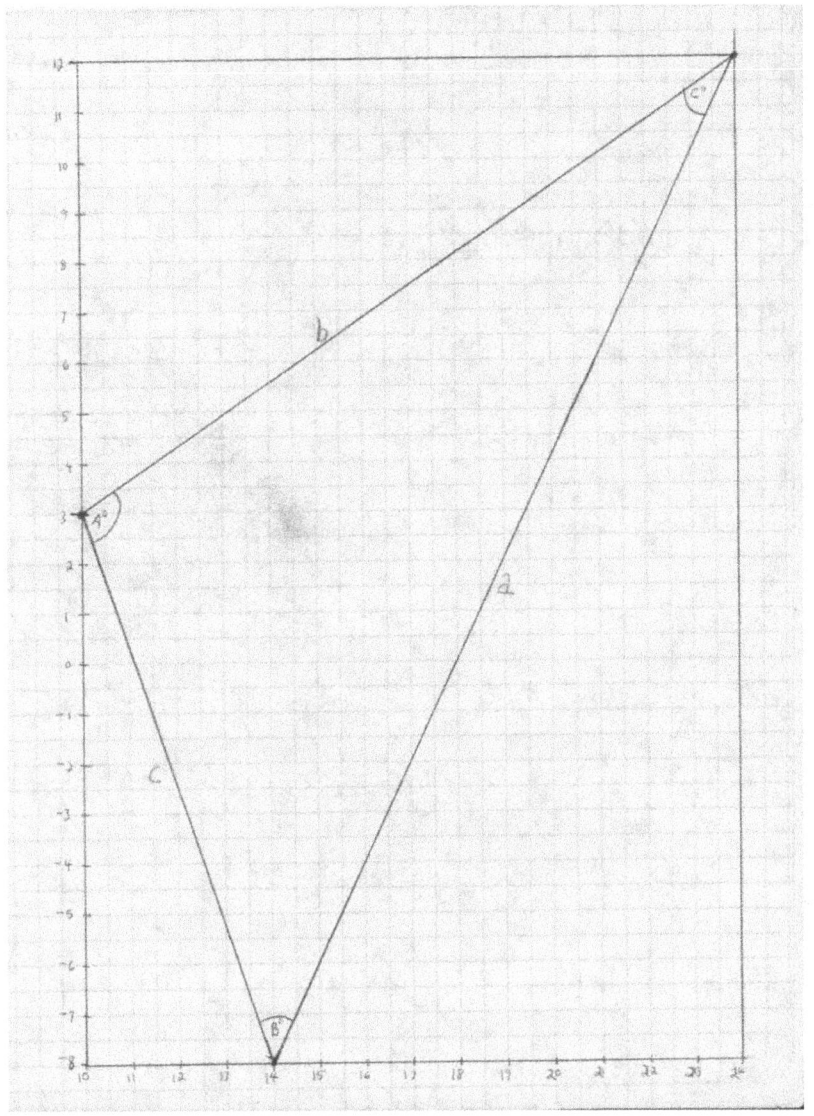

48.) Find the length of the sides of the triangle.
a =22.36067977 b.)16.64331698   c.)   11.70469991

49.) The area of the triangle = 18.86676236 because
[(25.35434833)(25.35434833-22.36067977)
(25.35434833-16.64331698)(25.35434833-11.70469991)]$^{0.5}$ =
[25.35434833)(2.99366856)(8.71103135)(13.64964842)]$^{0.5}$ =
                        (355.9547218)$^{0.5}$ = 95.00000003 = 95

50.) Ln of 45/Ln 30= 3.80666249/3.401197382

51.) In the graph, where are the two places where the function, 3X, meets the other function $16X^2-9X-4$?

$16X^2-9X \quad -4 = 3X$

$16X^2-12X-4 = 0$ and after diving by 4 is $4X^2-3X-1 = 0$

$3 +/- (9 + 16)^{0.5}/8$ gives two X values, 1 and -0.25. Placing these into the function, $16X^2-9X-4$, give us the X, Y results of (1, 3) and (-0.25, -0.75).

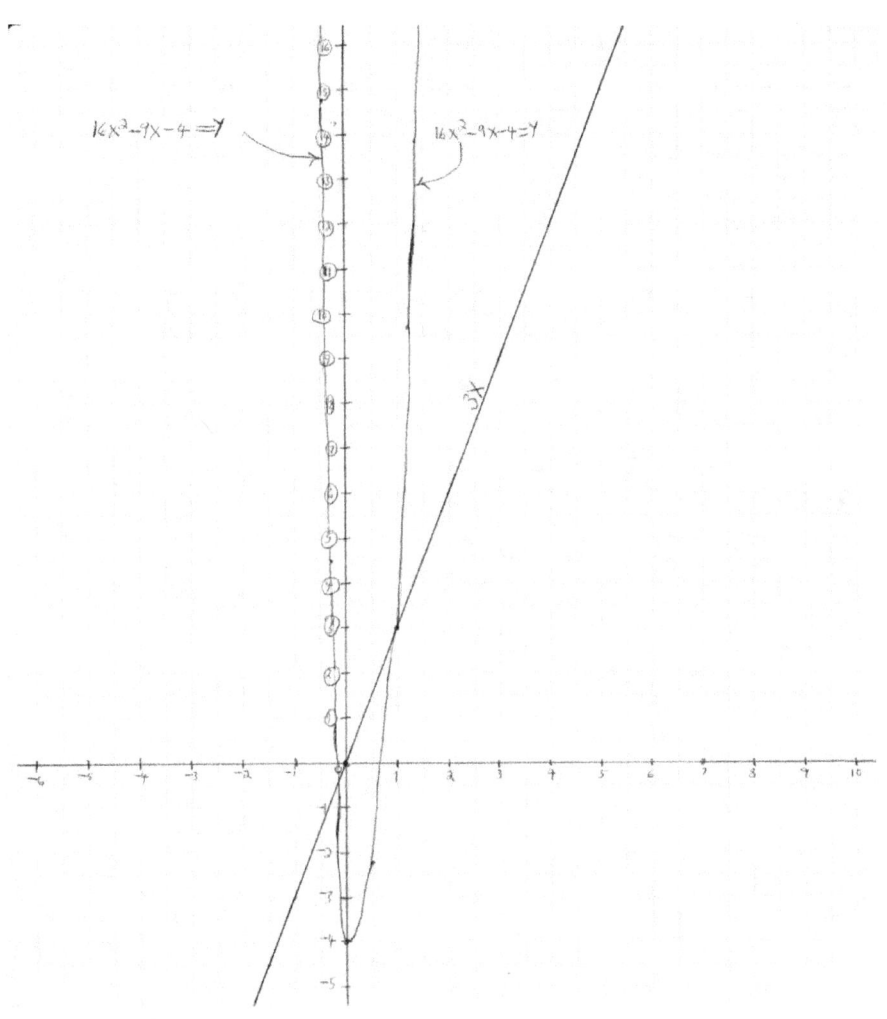

16x² - 9x - 4 = Y

16x² - 9x - 4 = Y

3X

52.) What is the distance between points 66X, 3Y and 15X, -18Y?
(To solve this you do not need to have a graph.)

$$[(66-15)^2 + (3-18)^2]^{0.5} = [(51)^2 + (-15)^2]^{0.5} = [2,601 + 225]^{0.5} = 53.1601$$

53.) Find the distance between the circumferences of the two
      following circles. The top circle has a radius of
      5.852349955 yards and the inside triangle reaches the,
      within a circle, x axis at zero and at 5.5. Its origin is
      at(7, 16). The bottom circle has a radius of
      5.656854249 and the inside triangle reaches the x axis at
      zero and four. Its origin is at (-10, -8.5). Also, what
      is the purpose of reporting the coordinates of the
      triangles that are inside of the circles?

$$[(7 - -10)^2 + (16 - -8.5)^2]^{0.5} = (889.25)^{0.5} -$$
$$(5.852349955 + 5.656854249) = 18.3110909.$$

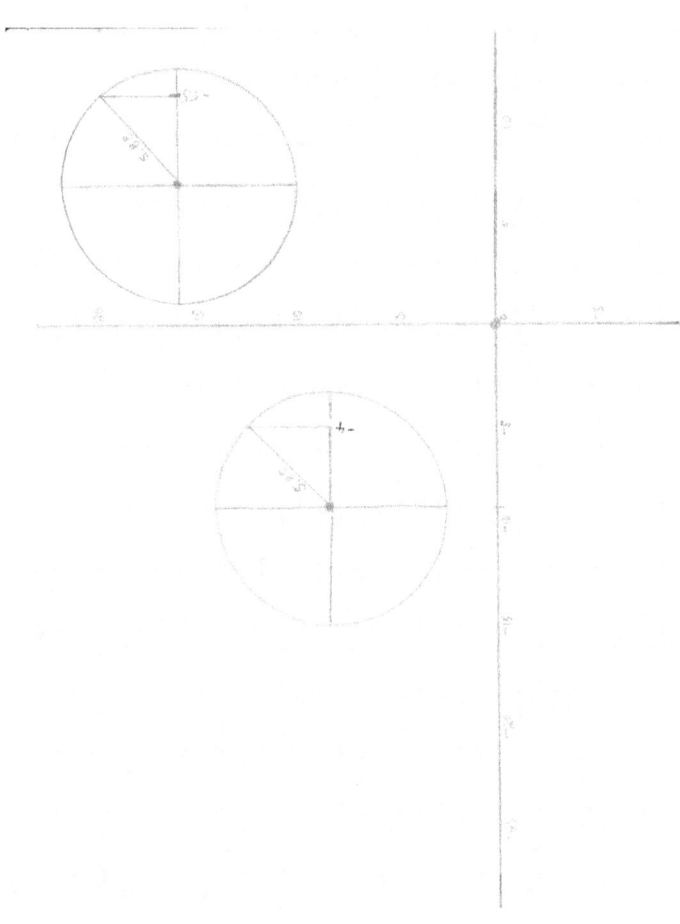

In the above graph the vertical line is the X axis and the horizontal line is the Y axis.

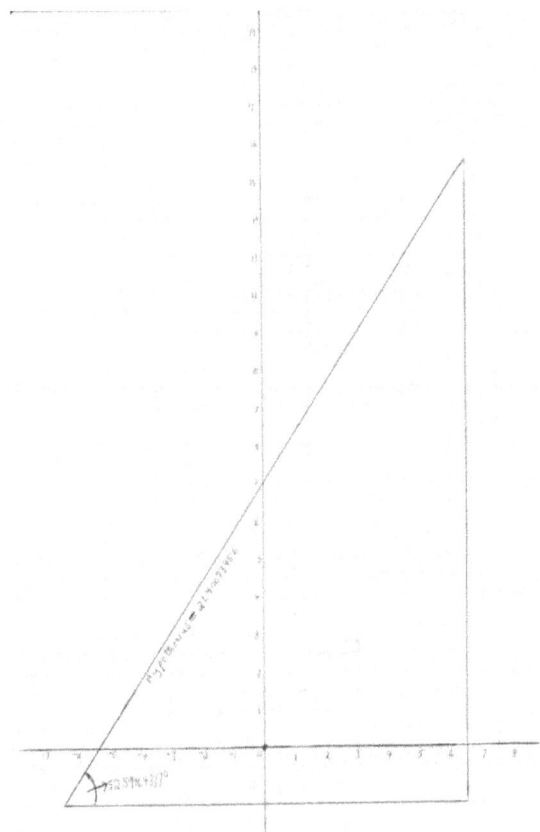

54.)   The bottom left angle of the above triangle is 52.59464337°. What times the bottom left angle's degrees equals the length of the triangle's X axis and what times the bottom left angle' degrees equals the length of the triangle's Y axis?   The left side corner of the triangle is of (-6.5, -1.5), the bottom right corner of the triangle has a location of (6.5, -1.5), and the top right location of this triangle has a location of (6.5, 15.5).

Multiplying the hypotenuse, 21.40093456, by the cosine of the angle will give you the length of the X which is 13. To find the Y axis of the triangle, multiply the hypotenuse, 21.40093456, by the sine of the angle that has a measurement of 52.5946337°. The sine of this angle is 0.79435773 and multiplying this result by the hypotenuse of the triangle will give you the length of the triangle's Y axis as being seventeen, 17.

Also, the arc Tangent, $\tan^{-1}$, of the division of the length of Y divided by the length of X, the arc $\cos^{-1}$, the division of the length of Y divided by the length of X, and the arc $\sin^{-1}$, the division of the length of Y divided by the length of X, all give a result of 52.5946338° for the bottom left angle of the triangle.

55.) $16X^2 - 9X - 4 = 3X$, solve for X.

$16X^2 - 12X - 4 = 0$

$4X^2 - 3X - 1 = 0$

$(4X+1)(X-1) = 0$ So, here, X must equal negative one quarter or one.

___

56.)   To answer the first question we can use any of the four information parts at the top of the page that do not have a number equal to zero incidents. Let's uses the fourth item, assaults.

$$\frac{609.9}{100,000}(X) = 5 \text{ Assaults}$$

To solve for X, five assaults should be divided by 609.9 and then multiplied by 100,000. Here, it can be found that X equals 820.0082001 or 820 people and that Autaugaville, Alabama had a population of eight hundred and twenty people in the year 2001.

To answer the second question we need to multiply Imaginationville's 2009 population by eight and one-half percent, 0.085.

$$(0.085)(18, 902) = 1,606.67$$

This is the number of births of that year. To find the number of births per one hundred thousand people we can use Y as the unknown per one hundred thousand people. Then we can multiply that fraction, the unknown per one hundred thousand people, by the population of Imaginationville before the births appeared.

$$\frac{Y}{100,000} \cdot (18,902) = 1,606.67$$

To solve for Y we will multiply 1, 606.67 by 100, 000 and then divide the result by 18, 902. This will show Y to equal 8,500. Again, because the birth rate was eight and one-half percent then when that growth rate is multiplied by one hundred thousand people the result must be eight thousand five hundred people. So, Y equals eight thousand five hundred.

57.)

Ms. Bart was approved to receive a fifty-two thousand two hundred dollars mortgage. The interest rate on the mortgage is seven percent and she has twenty-four years to pay the mortgage. Her house will have a monthly cost that will include taxes and other costs that are two thousand seven hundred and eighty-six dollars and twenty-eight cents plus the amortization of the fifty-two thousand two hundred dollars financing of the cost of the house that she purchased after paying a ten percent down payment on the sale of the house.

What was the sale price of the house before she made any down payment?

To find the quantity that was present before a discount was used we should divide the amount that remains after the down payment by the sum of one minus the down payment.

$$\frac{52,200}{(1-0.1)} = 58,000$$

58.) What will be her monthly amortized mortgage payments?

$$\frac{X[(1+ I/12)^{n+1}}{I/12} = \frac{X[(1+ 0.07/12)^{288+1} -1]}{0.07/12}$$

X(749.2418193)   = $52,200

X   = $69.67 will be the amount taken off of the mortgaged amount during the first payment.

$$\frac{52,000(0.07)}{12}$$ = $304.5 is the interest cost during the first month.

The amortized monthly mortgage payments will be $374.17

59.) In order to pay this mortgage within twelve years how much extra money must she pay monthly?

$$\frac{X[(1+ I/12)^{n+1}}{I/12} = \frac{X[(1+ 0.07/12)^{144+1} -1]}{0.07/12} \quad =$$

$$\frac{X(1.324199947)}{0.00583333} = \$52,200$$

So, X   equals $229.95 and the first month's interest payment will be $304.5. The total monthly amortized mortgage payments will be $534.45 and the additional payments, compared to the minimal monthly payment of her mortgage is 160.28

This is a savings of 288/2 months times $374.17 or $53,880.48.

California schools are facing two issues regarding textbook acquisition: availability and cost. More than half a million students do not have textbooks to use in class and approximately two million students cannot take textbooks home to do homework.[1] School textbook prices have risen alarmingly in recent years. Depending on the subject, a single elementary textbook can range in price from $30 to $100. Legislation should be enacted to reduce the cost of K-12 school textbooks.

60.) Triangle D has Point A which is at (-16, 8), Point B which is at (21, -1), and Point C which is at (13, 0). Across from Point A is side a, across from Point B is side b, and across from Point C is side c. Find the degrees of these three points of Triangle C.

By observing the locations of the three corners of the triangle we can find the distance between the corners which is the length of - triangle sides a, b, and c.

Side a is across from Point A, side b is across from Point B, and side c is across from Point C. Because Point A is at (-16, 8), Point B is at (21, -1) and Point C is at (13, 0) we can find the distance between Point A and Point B to be across from Point C, so it is side c. The size of this distance between Point A and Point B is

$$[(-16 - 21)^2 + (8 - -1)^2]^{0.5}$$ is 38.07886553. This is side c.

The distance between Point B and Point C, which is directly across from Point A and it is side a, is found by using subtraction and the distance formula. Side a is found by using above following facts.

$$[(21-13)^2 + (0 - -1)^2]^{0.5}$$ or 8.062257748.

Last, the distance between Point C and Point A is

$$[(113- - 16)^2 + (0 - 8)^2]^{0.5}$$ or 30.08321791. It is side b.

From the information above we can use the Law of cosines to find the three angles of Triangle C.

The Law of Cosines

a^2 $= b^2 + c^2 - 2bc\cos A$

$2bc\cos A$ $= b^2 + c^2 - a^2$

$\cos A$ $= \dfrac{b^2 + c^2 - a^2}{2bc} = \dfrac{30.08321791^2 + 38.07886553^2 - 8.062257748^2}{2(38.07886553)(\_30.08321791)}$

So, $\cos A$ $= \dfrac{30.08321791^2 + 38.07886553^2 - 8.062257748^2}{2(38.07886553)(\_30.08321791)}$

$2{,}290 / 2{,}291.069619$ $=$ 0.999533136

A $=$ $\cos^{-1}(0.999533136) = 1.75.0853148°.$

Here, use of the parentheses, from your calculator, after the term $\cos^{-1}$ is very important.

---

b^2 $=$ $a^2 + c^2 - 2ac\cos B$

$2ac\cos B$ $=$ $a^2 + c^2 - b^2$

$\cos B$ $=$ $\dfrac{a^2 + c^2 - b^2}{2ac}$

So, $\cos B$ $= \dfrac{(8.062257748)^2 + 38.07886553^2 - 30.083211791^2)}{2(8.062257748)(38.07886553)}$

$= 0.993480072$

B $= \cos^{-1}(0.993480072) = 6.546290545 °$

$$c^2 = a^2 + b^2 - 2ab\cos C$$

$$2ab\cos C = a^2 + b^2 - c^2$$

$$\cos C = \frac{a^2 + b^2 - c^2}{2ab}$$

So, $\cos C = \dfrac{(8.062257748^2 - 30.08321791^2 - 38.07886553^2)}{2(8.062257748)(30.08321791)}$

$$C = \cos^{-1}(0.989532982) = 171.7028554°$$

Angle A   =   1.750853148°

Angle B   =   6.546290545°

Angle C   =   +171.702855400°
179.9999991   This is very close to 180°.

Sarah E Goode invented an improved cabinet bed and received patent 322,177 on 7/14/1885. She (b.1850s-d.c. 1905) was the first African-American woman to receive a patent from the United States Patent and Trademark Office. However this claim is disputed by some that believe **Marjorie Joyner** to be the first African-American woman to receive a patent. This has been taken from Wikipedia, the free encyclopedia.

Here are three important mathematics things to remember;

A.) PEDMAS   B.) FOIL &   C.) The Binomial Theorem

From the University of New Hampshire

UNH currently states that the expectation is that students will study (on average) 2-3 hours per week outside of class for every hour spent inside class. This expectation is communicated to new students at both the June and the September orientations and is provided to those students in a letter in June. This standard of expectation is widespread among other universities.

I went to a Philadelphia School Board meeting in a church on Broad Street and asked how many hours a week should students study. I received the same answer as what UNH expects of its students. This equals studying for about thirty, or more, hours per a week.

South Carolina has the Highlands and the Lowlands. So, bicycling from the northwest section of the state to the southeast section of the state is like falling down a mountain, in other words the bicycle and you will "MOVE FAST, VERY FAST."

During a long distance bicycle trip, I visited a bicycle shop for help. The worker who helped me told me just what I told him. That was that the hills of South Carolina can be deceiving. There they do not allow you to stay on the correct side of the road and then, often, you will find out that you are on the wrong side of the road when a large tractor-trailer is heading towards your bicycle from the front of you. I was happy to find someone who had found the same difficulties that I found. He told me that the roads are dangerous in the hills because they can deceive you, much.

That year I arrived there and went to the Olotunji Village of Sheldon, South Carolina. There, Baba Obuyo showed me the village, after the village had stopped the visitor's hours. They grow foods there and they have a good school for their children. The man who started the village, Adefunmi, and the members of the village are people who practice the Yoruba way of life. This is also the religion that I practice. Baba Obuyo showed me a wall of the village that had my names, Olaniyan Adefunmi being one, on it. The names were spelled in a few different phonetic ways. He told me that my, name Olaniyan, represents the fourth grandson of a grandmother. My mother had three sons and none of her siblings had any male children. One of her children died and did not live long enough to be around when I was born. Many years earlier a spiritual reader, who did not know me, told my sister that the child who died was a boy even though many years before then my mother told me that doctors did not tell her the sex of the child. If it was a boy then I am my grandmother's fourth grandson and my name means exactly what   Baba Obuyo told me that is means.

Obuyo at the Oyotunji Village

And all that Jazz at the

. <u>African American Museum</u> .

One day, less than one hundred years ago, or so, I sat in the Afro-American Museum in Philadelphia, Pennsylvania. There, we were to explore the sounds of, and be explored by the sounds of, jazz. Only those chosen people who can remove all unnecessary smiles, etc., can join a "real intellectual" audience such as the audience that included me. You know what I'm sayin."

The progressive jazz moved deeper than just our ear drums and the harmony was superb, or something.

Then part two arrived. This included a solo horn player's music. This was some "deep stuff" so we listened, intensely. In front of the row of chairs that included me was a small family that had a little girl who was about five to eight years old. She was not there to tell us that she has "been there and done that" and the sound of intellectual jazz progressing through the air was an abstract substance for her not to enjoy, in her opinion. The musician did not stop there but he made awkward screams with his horn's notes and chose quiet time when it was least expected which allowed him to increase the volume at an adjacent time period which started with a loud trembling loud volume of a chard that no human has ever heard. This caused the little girl to jump in her seat. Thank God that her movements could not increase her volume or height significantly because she was very small, in the first place. Yet, the more the man with the long beard and very long and thick hair locks introduced and tried to scare those who did not like his music with minor and augmented scales she moved in combating weird configurations to his music. When he went up she jumped down and when he went down she jumped up except for after he had quieted down much and after he decided to reintroduce his "noise" she would repeat with more vigor a response against every mistake he could create with the horn. Her mother did not notice her daughter's behavior. I thought that she had the right to express her feelings about the abstract jazz.

4n 83948082b8 349039024--------------------------------2489042389298 ---------
--------------2u9 0499\\\\\\\\\\\\\\\\\\\\4 890_____ 8248 4880

O.K. I was for the Patriots, the football team, to win the Super bowl of 2009 not only because a truck smashed into my head in Baton Rouge, Louisiana but because I thought the team to be "all that," or something similar to that.

After they won I went on and went to bed. There I had a dream of the times when France had colonies both in Louisiana and Haiti. Haiti, unlike what some believe, is not an island but shares an island with a nation named the Dominican Republic. The nation is only about the size of the state of New Jersey. Haiti fought for her independence from France and won in 1804 under leader ship of people like Jean Jacques Dessalines. Today, Haiti is very poor and many blame her poverty on the fact that she removed France and has many citizens who practice Vodun, a religion that many incorrectly name as being Voodoo. "Get Real," Voodoo comes from Hollywood and is not like Vodun, a

In that dream I wondered if there was a large mathematics discovery that occurred in Haiti. The next morning I rushed to Temple University to find who discovered something mathematical and who was a native Haitian. I found his name, Leon Romain. He solved an over five thousand years old mathematics problem that many believed could not be solved. Since then he wrote a book about his discovery I have been trying to locate him so that I could learn what he has discovered.

Leon Romain has devised a theorem for trisecting any angle, one of geometry's great puzzles. If he is right, it could change your life.

HAITI

249n u240ut2429t581249-85400000000000000000000000000013bn 82498b
n8_____89340u9234u090u4204ut9 9240 204\\\\\\2 4

The information below is a collection of attempts to reach Leon Romain. He has discovered, unless he is proven to be wrong, a mathematical "break through." I like him to explain to me, us, in layman's terms what he has done. Yet, he does not have an email account that Temple University Library's email machines can reach. The publisher, editor@haitiprogress.com, is just as difficult to reach. Maybe one or more of the readers of the Leon Williams journal know of a way to reach him. I am interested in sharing his work because his work is not just something that we, or he, has done but something that has been or is being done, today. The book that he has written is "Angular Unity: The Case of the Missing Theorem." His creation reminds me of the painful truth of the Po Tolo Stars.

The finding of these stars has often been created towards Greece who named them the Sirus Stars. Yet, it was the Dogan, from the Kemit and the Mali areas, who found these stars and named them Po Tolo. They, the Dogan, even found one of them that is invisible. Some, according to the late Ivan Sertima, wondered whether these people's ancestors arrived from outer space because they knew so much of the Po Tolo Stars. Deep!

Fw: Angular Unity: The Case of the Missing Theorem...
Fri, February 12, 2010 9:16:45 PM
From: Olaniyan Adefumi <oadefumi@yahoo.com> ...View Contact
To:

editor@haitiprogress.com

Haiti.doc (23KB)

_____

I have been trying to reach Leon Romain by email but all of my attempts to reach him have failed. So, I am forwarding the intended email that I tried to send to him to you. The message of the email to him is also a message to you from me, Olaniyan Adefumi, in representation of many concerned people of the United States of America.

From: Olaniyan Adefumi <oadefumi@yahoo.com>
To: leon@kafou.com
Sent: Fri, February 12, 2010 9:04:59 PM
Subject: Fw: Angular Unity: The Case of the Missing Theorem

po4u29 vu094uv09423u09v2u2094u 9vnuu9349unu 9u395
0924u0924u2094u209u3049u0 0vu404u40 4u204u

I have read very little about the book that you wrote, Angular Unity: The Case of the Missing Theorem. I know little of its materials, yet I do like mathematics, at least the parts of it that I do understand, much. I write for a local journal in Philadelphia, Pennsylvania, the Leon Williams Journal. Readers of the journal are very interested in how communities of "third-World" places, like Haiti, are doing economically, culturally, and in other manners. In the journal I focus, monthly, on issues of mathematics.

After the United States of America football game, not a soccer game, that is known as the Super bowl was won by the team that I wanted to win, since the team that represents Philadelphia, Pennsylvania, the City of where I live and that is represented by the football team named the Eagles, lost a chance to play in the Super bowl, and after the most recent earthquake occurred in Port-au-Prince, I decided to go to the Internet and researched for mathematics discoveries from Haiti, a place that has a relationship to France just like the city of New Orleans, the city that has the team, the Saints, that received a victory in the Super bowl football game has an historic relationship to France.

I appreciate if you would give to us, in layman's terms, what you have discover, in mathematics, and what was believed for many years, in relationship to what you discovered, that could not be done. Then you did it.

If you have questions from school, etc., send them to me, Olaniyan Adefumi, oadefumi@yahoo.com. Questions will be answered, shortly.

Your note books are sacred. One note book can be used in class for all subjects. At home, the library, or elsewhere, transfer your notes form this notebook to notebooks that are specifically for only one subject. Review your notes, often. Also, carry index cards that have some of your notebook information. Study these cards when you are on the bus, train, etc. Repetition of facts makes learned materials part of your mental assets.

Hesabu (Swahili for Mathematics) puzzle **Three Three-Digit Numbers**

(Here is a task for people of the seventh grade, or anyone else who would step up to the challenge of a puzzle.)

Find three sets of three-digit positive numbers where the first two of these numbers sum to the third three-digit number and no single digit is to be used more than once.

Here is an example from a seventh grade student of Jones Middle School of Philadelphia, Pennsylvania.

$$734$$
$$\underline{216}$$
$$950$$

Here, no single digit is used more than once, all of the rows of numbers have three digits, and the first two rows sum to the third row of numbers.

The game can be expanded beyond addition. An example of how it can be used with subtraction is in the following example.

$$467$$
$$-358$$
$$\overline{}$$
$$109$$

Here, the result of 467 minus 358 is 109. Again, no single digit has been used more than once and the difference between the two stating numbers, of rows one and two, gives the result of row three.

The game can be expanded to the requirement of each digit alternating from being odd to being even. Here, 129 + 438 equals 567.

Assignment: Find two examples of addition and subtraction that are limited by the rules of this game, Hesabu.

The Trip to Vulture Hill

You can leave Valleytown anytime for the three hundred mile trip to Vulture Hill and you must be there before Monday at eight-thirty-seven in the morning. There is one road, Vulture Valley Road, which you can use to get there. You need one hour and a quarter of an hour to prepare for a job interview in Vulture Hill.

Now, it is eleven-thirty-seven in the morning of Friday. You must go to Vulture Hill by foot or by bicycle. You can walking or jogging, with your materials, at a two and forty-sixth hundredths of a mile per an hour pace that will face similar reductions of speed like what you will experience should you bicycle instead.

From now until three-twenty p.m. it will be clear and sunny, yet, conditions will change. Every day, daylight will start after seven-five in the morning until it is nine-ten in the evening. At that point travelling speed by bicycle will be limited to being only sixty percent of the normal speeds.

The temperature will reach a high of one hundred ten degrees from eleven a.m. to two p.m. Once the temperature is greater than eighty-five degrees your bicycling speed will be reduced to being only eighty-eight and one-half percent of your normal maximum speeds of fifteen miles per an hour. This speed is a result of the need to carry many personal items within your backpack and in your bicycle's basket. This reduction of speed will occur within every hour of continuous bicycling when the temperature is greater than eighty-five degrees. The temperature will be sixty degrees at six a.m. and it will get to be ten degrees warmer every hour during the morning until the eleven a.m. maximum is reached where it will fall by ten degrees during additional hours until it reaches being only sixty-five degrees, again. The change in temperatures will be gradual during every second of an hour.

Heavy rain will start on Sunday at ten thirty-three a.m. until seven-twelve a.m. on Monday morning and at eight a.m. until 9:38 during Monday morning. Under this condition your bicycling speed is reduced by sixty-three and one-thirty percent.

Turkey vultures will follow you because they sense the foods of your backpack and slow your speed by interfering with you traveling abilities between the two hundred and seventy-second mile marker and the two hundred and ninety-eighth mile marker. This will cause you to bicycle faster but by stopping to scare away the loud and annoying birds you will not progress through more distance than you would had if the large birds were not there.

A few trains go through this location. A slow moving Vulture Valley Railroad freight train leaves both cities to reach the other city seven and one-half hours later starting at 3:54 in the morning and thirty hours later and before the next 3:54 train is to start. It takes ten minutes to cross the intersections with Valleytown-vulture Hill Road. It uses a set of train tracks that the passenger trains do not use. It does not leave either city between the hours of mid-night and 3:54 in the morning. A passenger train will start from Valleytown, along the only two way passenger train route between Valleytown and Vulture Hill, every six hours and return after spending three minutes in Valleytown and in Vulture Hill. It travels at eighty-seven and one-half miles per an hour between each main road mark intersection except for between the one hundred and seventy-fifth and the two hundred and tenth train intersection with the main road where it moves at a rate of ninety-five and one over two and two tenths miles per an hour. It takes twenty minutes to cross each main road intersection, starting at the one-hundred and fifth mile marker. Each train track intersection is separated by thirty-five miles starting at the one hundred and fifth mile to the two hundred and eighty mile marker. All of these intersections are with the same road, Valleytown-Vulture Hill Road. A gate closes off the road every time a train is five minutes away from the road until the train has passed the main road for one minute.

At 12:22 p.m. on Saturday your tires get to be flat and to fix a flat takes fifteen minutes. Also, the road is very damaged, steep, and hard to use from the eighty-first to the one hundred and ninety-fifth mile marker. This condition slows down your speed by fifty percent plus it makes your back tire go flat.

Your travel speeds are predicated by the amount of sleep you had within the last twenty-four hours, the number of meals that you ate in the same period of time, whether you showered or bathed within the last twenty-four hours, the night where after sundown your speed is reduced by forty percent, your travel speed is only one-half the normal speed if you do not sleep for eight hours in one day and your speed will be reduced by ten percent for every waking hour during the twenty-four hour period that you did not get eight hours of sleep. So, if you got seven hours of sleep the result will be the same as if you did not get any hours of sleep. You do not have to sleep in a room on a bed and you can sleep outside but there sleep hours will be multiplied by one-half. So, to get the correct number of sleep hours you will have to spend much more time trying to sleep than you would have to spend doing the same thing inside a bedroom. Plus, the condition will be the same if you did not shower or bathe plus ate three meals that are separated by at least six hours each within the last twenty-four hours. Travel from mid-night to five a.m. can be done no faster than seventy-five percent of what you otherwise normal traveling speed would be. Each meal takes one hour of your time and taking a shower or a bath takes one-half hour. At locations ten miles after the train tracks there are one or more all-night restaurants, a motel or motels, and a place, such as gas stations, where you can use a toilet. You have foods within your backpack. It takes one and one-half hours to wake and eat and prepare for more traveling.

If you got there on time, what "short-cuts," if you used any, did you take? Over the radio you heard an announcement during Monday morning's seven o'clock morning news that because of the heavy rains many businesses and schools will open six hours later than their usual starting time.

When will you arrive at Vulture Hill? The Trip to Vulture Hill

Here is one, not the only, explanation of how one could finish the trip to the job interviewing firm.

Time	Activity	Conditions	Mileage & Mathematics	Cumulative mileage
Friday 11:37 A.M.	Eating	The temperature is above 110 degrees.		0 Miles
12:37 P.M.	Bathing			0 Miles
1:07	Bicycling		(15M/hr.)(3.383333 hrs.)(0.885 because of the temperature is above 85°degrees) = 44.91375 Miles.	44.91375 Miles
4:30	Bicycling		(15 M/hr.) = 15 Miles.	59.91375 Miles
5:30	Eating			
6:30	Bicycling	Hills, etc.	(15M/hr.)(1.40575) = 21.08625 Miles.	81 Miles

Time	Activity	Conditions	Mileage & Mathematics	Cumulative mile-age
7:54.345	Bicycling		(15M/hr.)(0.5)(1.2 hours) = 9 Miles.	90
9:06	Fix Flat Tire.	There is a damaged Road from mile 81-195.	0 miles	90 Miles
9:21	Bicycling	Night started at 9:10	(15M/hr.)(0.6)(0.5)(2.15 hours) = 9.675 Miles.	99.675 Miles
11:30 P.M.	Eating		0 Miles	99.675 Miles
Saturday 12:00 A.M.				

Time	Activity	Conditions	Mileage & Mathematics	Cumulative mile-age
12:30	Biking		(15M/hr.)(0.6)(0.5)(4.5407) (0.75) = 15.325 Miles.	115 Miles
05:00	Bed			
11:00	Wake-Eating			
2:30	Bicycling		(15M/hr.)(0.885)(0.5) =6.6375 Miles	121.6375 Miles
3:30	Bicycling		(15M/hr.)(0.885)(0.5)= 6.6375 Miles	128.275 Miles
4:30	Bicycling		(15M/hr.)(0.5)(2.5 hours) = 18.75 Miles	147.025 Miles
7:00	Eating			
8:00	Bicycling		(15M/hr.)(0.5)(0.396666)= 2.975 Miles	150 Miles
8:24		Motel		
9:00	Bath			
9:30	Leisure			
11:00	Bed			

Time	Activity	Conditions	Mileage & Mathematics	Cumulative mile-age
Sunday 7:00 A.M.	Wake-up and Eat			
8:30	Bicycling		(15M/hr.)(0.885 Heat)(2.05time) (0.5damaged road) = 13.606875 Miles	163.606875 Miles
10:33		Heavy Rains	(15M/hr.)(0.885Heat) (0.36666Rain)(0.5)Bad rd. conditions)(2.95hrs.) = 7.179562487 Miles	170.7864375 Miles
1:30 P.M.	Eating			
2:30 P.M. Sunday	Bicycling		(15M/hr.)(0.885 Heat) (0.36666 Rain)(0.5 Bad Road Conditions)(2 Hours) = 4.8675 Miles	175.6539375 Miles
4:30	Bicycling		(15M/hr.)(0.36666 Rain) (3 Hours)(0.5 Damaged Road) = 15 Miles	190.6539375 Miles

Time	Activity	Conditions	Mileage & Mathematics	Cumulative mile-age
7:30	Eating			
8:30	Bicycling		(15M/hr.)(0.36666Rain) (0.5 Damaged Road)(3.5 Hours) = 9.625 Miles	200.2789375
Mon-day 12:00 A.M.	Eating			
1:00	Bicycling		(15M/Hr.)(0.36666 Rain)(0.5 Damaged Road)(0.75 After Mid Night)(0.116666 hours) = 0.002475	200.5195568
1.07 A.M.	Bicycling		(15M/Hr.)(0.36666 Rain)(0.75After Mid Night)(4.883333Hours) = 20.14375 Miles	220.6633068
6:00	Eating			
7:00	Bicycling		(15M/Hr)(0.2Hours)(0.36666Rai) (0.9No Sleep) =0.99 Miles	221.6533068 Miles

Time	Activity	Conditions	Mileage & Mathematics	Cumulative mileage
7:12	Bicycling		(15M/Hr.)(0.43333Hours) (0.9 No Sleep) = 5.849955 Miles	227.5032618 Miles
7:38	Call the Interviewing Organization	Motel	(15M/Hr.)(0.9 No Sleep)(0.2 Hrs.) = 2.7 Miles	230.2032618 Miles
7:50	Sleeping			
9:00	Call the Interviewing Organization			
9:10	Sleeping			
4:00 P.M.	Waking-Up and Eating			
5:30	Bicycling		(15M/Hr.)(3.6666 Hours) = 55 Miles	285.2032618
9:10	Bicycling		(15M/Hr.)(0.6NightSpeed)(0.83333Hours)= 7.5 Miles	292.7032618 Miles
10:00	Eating			

Time	Activity	Conditions	Mileage & Mathematics	Cumulative mile-age
11:00	Bicycling		(15M/Hr.)(0.6NightSpeed) (0.810748689 Hours) = 7.2967382 Miles.	300 Miles
11:50	Prepare for Interview.			
12:50	You are at the Interviewing Firm.			

Is the information that was given by this traveler correct? If not, or if so, why?

    At Cheyney University I made a presentation of this information at school and a teacher was asked if I could publish it. He said that he will try to get it published. After graduating, I approached the University of Pennsylvania to find out whether anyone accomplished the mathematics that I had in my presentation. There, a teacher told me that he teaches his class the information that I had on parabolas. Then, he gave to me a book on calculus. This was good because I lost all of my calculus books. Going through the book where it has information on parabolas, I found nothing that overlapped what I found about parabolas, so far.

A Structure in side the country of Germany   .

Math Education in the U.S., Germany, and Japan: What Can We Learn from This?

A recent study compared the videotaped teaching styles of 81 eighth-grade math teachers in the U.S. with those of teachers in Germany and Japan. What did educators learn from the study?

An April 27 Education World article (Math Wars!) reported that U.S. eighth graders scored below average in math on The Third International Mathematics and Science Study (TIMSS). In the study, the most comprehensive international comparison of math and science achievement levels ever attempted, U.S. math students were outperformed by students in the countries of some of our closest economic allies -- and major economic rivals. Many people -- including parents, politicians, educators, and business leaders -- wanted to know why. The results of another component of TIMSS have been released -- and they may help answer the question.

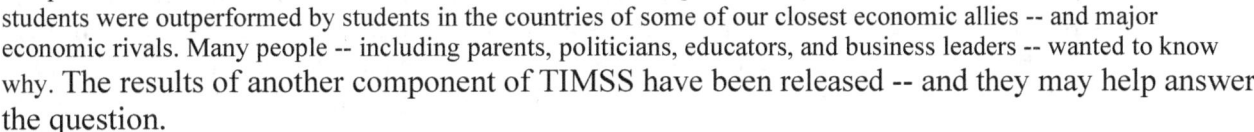

POSING THE QUESTION

In a cross-cultural study of educational philosophy and teaching style, researchers at the University of California (Los Angeles) compared the instructional methods of 8th grade math teachers in the U.S. with those of teachers in the countries of two of its most important economic competitors -- Germany and Japan. The study had among its goals to:

- learn how 8th grade mathematics is taught in the United States, Germany, and Japan.
- develop objective measures for evaluating classroom instruction.
- determine how teaching methods in the three countries conform with current U.S. reform recommendations.
- determine U.S. teachers' perceptions of current reform recommendations.

To accomplish those goals, the UCLA researchers videotaped 231 eighth grade math lessons -- 100 in Germany, 50 in Japan, and 81 in the United States -- as they were being taught. The researchers then spent months viewing, analyzing, and discussing those lessons. The results of The TIMSS Videotape Classroom Study revealed striking differences in the teaching styles and educational focus of math teachers in the United States and of their German and Japanese counterparts.

SEARCHING FOR SOLUTIONS

According to the research team, the videotapes revealed major international differences in:

1. **Lesson goals**--German and U.S. teachers stressed the development of problem-solving skills, while Japanese teachers stressed students' understanding of underlying concepts. More than 60 percent of U.S. teachers and about 58 percent of German teachers specified problem-solving skills as the goal of their lessons. More than 90 percent of Japanese teachers emphasized conceptual understanding over problem-solving ability.

2. **Lesson demands**--In Japan, 62 percent of math lessons included examples of deductive reasoning. Only 21 percent of German lessons and 0 percent of U.S. lessons required deductive reasoning. Deductive reasoning is defined as the reasoning needed to draw logical conclusions from premises.

3. **Lesson difficulty**--Topics covered in U.S. 8th grade classrooms were judged to be at a 7th grade level according to international standards. Topics covered in Germany were at an 8th grade level and topics taught in Japan were determined to be at a 9th grade level.

4. **Lesson focus**--Math lessons in Japan appeared more specific and coherent than U.S. math lessons. In Japan, lessons focused tightly on a single mathematical concept and teachers provided clear connections between different parts of the lesson. U.S. math lessons contained significantly more topics than Japanese math lessons; U.S. teachers switched from one topic to another with greater frequency; and U.S. teachers were less likely to provide explicit links between topics.

5. **Lesson content**--In the United States and Germany, about 90 percent of student seatwork involved practicing routine procedures. In Japan, 41 percent of working time was spent on routine practice and nearly half the time was spent "inventing new solutions and engaging in conceptual thinking."

6. **Lesson organization**--U.S. and German math lessons generally had two separate steps or phases. The first was the acquisition phase, in which the teacher demonstrated how to solve a problem. That was followed by the application phase, in which students practiced solving sample problems while the teacher helped individual students. In Japanese lessons, the procedure was reversed. Students began by solving a problem on their own, using information learned in previous lessons. They then shared their solutions and methods with one another and worked together to develop an understanding of the underlying concept.

7. **Lesson development**--In U.S. lessons, mathematical concepts and procedures were usually stated by the teacher. In German and Japanese lessons, concepts were generally developed through examples, demonstrations, and discussion.

8. **Lesson inviolability**--U.S. lessons were more likely to be interrupted by announcements and visitors than were Japanese or German lessons. Twenty-eight percent of the U.S. lessons were interrupted by outside events, compared to 13 percent of the German lessons and 0 percent of the Japanese lessons.

9. **Reform implementation**--In areas such as individual problem-solving, generating alternative solutions, and articulating conceptual understanding, Japanese teachers appeared more in line with the spirit of reform advocated in the U.S. than did U.S. teachers. Although most of the U.S. teachers believed their methods were consistent with reform recommendations, many still emphasized the "acquisition and application of skills" over conceptual understanding.

10. **Achievement level**--Japanese students scored among the highest in the world on the TIMSS. U.S. and German students scored about, or a little below, average.

EVALUATING SOLUTIONS

James W. Stigler, the UCLA psychology professor who directed the TIMSS classroom study, warns against drawing simple conclusions from these observations, however. In an article in Phi Delta Kappan, Understanding and Improving Classroom Mathematics Instruction: An Overview of the TIMSS Video Study, Dr. Stigler and co-author James Hiebert stress that teaching is a cultural activity that affects, and is affected by, a variety of social, economic, and political forces. One culture's educational system, however successful, can rarely be successfully imported into another culture, they say.

What we should learn from the Japanese educational system is not their style of teaching, but their approach to improving education through professional development. Japanese teachers, the authors point out, continuously participate in a formal process of collaboration and cooperation geared toward the refinement of individual lessons and the cumulative improvement of the educational process. No such organized approach to professional development exists in the U.S. "Our biggest long-term problem," the article states, "is not how we teach now but that we have no way of getting better."

As Dr. Stigler told Education World, U.S. educators focus too much on the teacher and not enough on the teaching. "We strive," he said, "to find the exemplary teacher when the real key is to improve the teaching of the average teacher." Dr. Stigler adds, "In this country, we need to develop a mechanism to improve teaching incrementally over time and we need to find a way to professionalize teaching by making professional development a part of every teacher's work week."

SHARING SOLUTIONS

The TIMSS Video Classroom Study has resulted in a number of recommendations intended to improve math instruction in this country by fostering opportunities for professional development. They include

- ensuring that teachers have a clear understanding of the spirit of recommended math reforms;
- providing beginning teachers with more concrete guidance and direction;
- assigning teachers lighter instructional loads; and
- providing teachers with opportunities to interact, discuss, share, and develop ideas and procedures for effective teaching.

For teachers themselves, however, perhaps the most useful result of the study is the availability of the TIMSS videotapes. These concrete instructional models provide teachers with the opportunity to view, assess, and compare alternative methods of teaching; to become aware of elements of their own teaching that may have become automatic and unquestioned; and to develop ways of improving the level of teaching in their own classrooms.

According to Dr. Stigler, "Efforts to improve student learning succeed or fail inside the classroom....We must study directly the processes that lead to learning in the classroom, for if we do not understand these processes, we will have little chance of improving them."

Parabola-Parabola Presentation

	The eight sections of finding the distance between two non-intersecting parabolas, even if one or both of the parabolas are not horizontal or vertical.
1	Make one parabola to be a vertical parabola.
2	Make a less flat vertical and positive parabola.
3	If a line, with an angle of 45 degrees or 135 degrees, can connect the two functions then the distance between the two parabolas might be found by sliding one or both of the two parabolas along that line until the two functions touch each other.
4	Otherwise, find the half-way point that is between the two vertices.
5	From the center of the Directrix line of the flatter parabola draw a new line that is parallel to the line that connects to both vertices and this new line should stop at a perpendicular point from where the half-way point between the two vertices has been found to be. Where this new line stops is the center of the desired circle.
6	Here, after finding the center of the circle, create a circle radius line, r, that will allow the circumference of the circle to be tangent to both parabola functions.
7	Connect the two points where the circle touches the two parabolas with a straight line. This line shows the distance between the two parabolas.
8	This technique of using a circle can be used to find the distance between two parabolas where one parabola is a vertical parabola and the other parabola is not, necessarily, a vertical or a horizontal parabola.

Finding the distance between two non-intersecting parabolas is the basis of the name and the purpose of this presentation. Triangles are what people observed and used to formulate trigonometry and observing principals of parabolas were used to form this presentation.	
Do you remember when your parent/s or guardian/s signed you up for your first pre-calculus class? Once you entered the class the first thing she or he wanted you to do was to maximize the fenced area of a yard with a limited amount of fencing?	(In economics the total revenue curves are often parabolas.)

A.) A square is a special type of rectangle.

B.) The yard is

maximized

by allowing X to

equal Y. This

helps us to find

the distance

between two

non-intersecting

parabolas about

* 66% of the time.

C.) The volume of a

cube can be

maximized by

letting the values

of X , Y, and Z be

equaled to each other.

Four Scenarios	

A.) The lease flat parabola's derivative equals 1.00 or -1.00 and a line with a slope of one or negative one extends from this location towards other parabola.
B.) Above the area where a line with a slope of one can reach the other parabola before it crossed the other parabola's axis of symmetry.
C.) Below the vertex of the flatter parabola
    1.) Problems,
        A.) increasing repetitive asymptotic conjugating issues.
 	
D.) Horizontal with vertical Parabolas	

Parabolas for Dummies
 	
    A.) Finding the Focus
 	
    B.) Finding the
 	
    Directrix	

Triangles are what people observed and what was used to formulate trigonometry. This plus the observation of principals of parabolas were used to form this presentation.	

Questions	
A.) Some answers might not exist, yet. This is a possibility since the only information of this topic that I could find.	

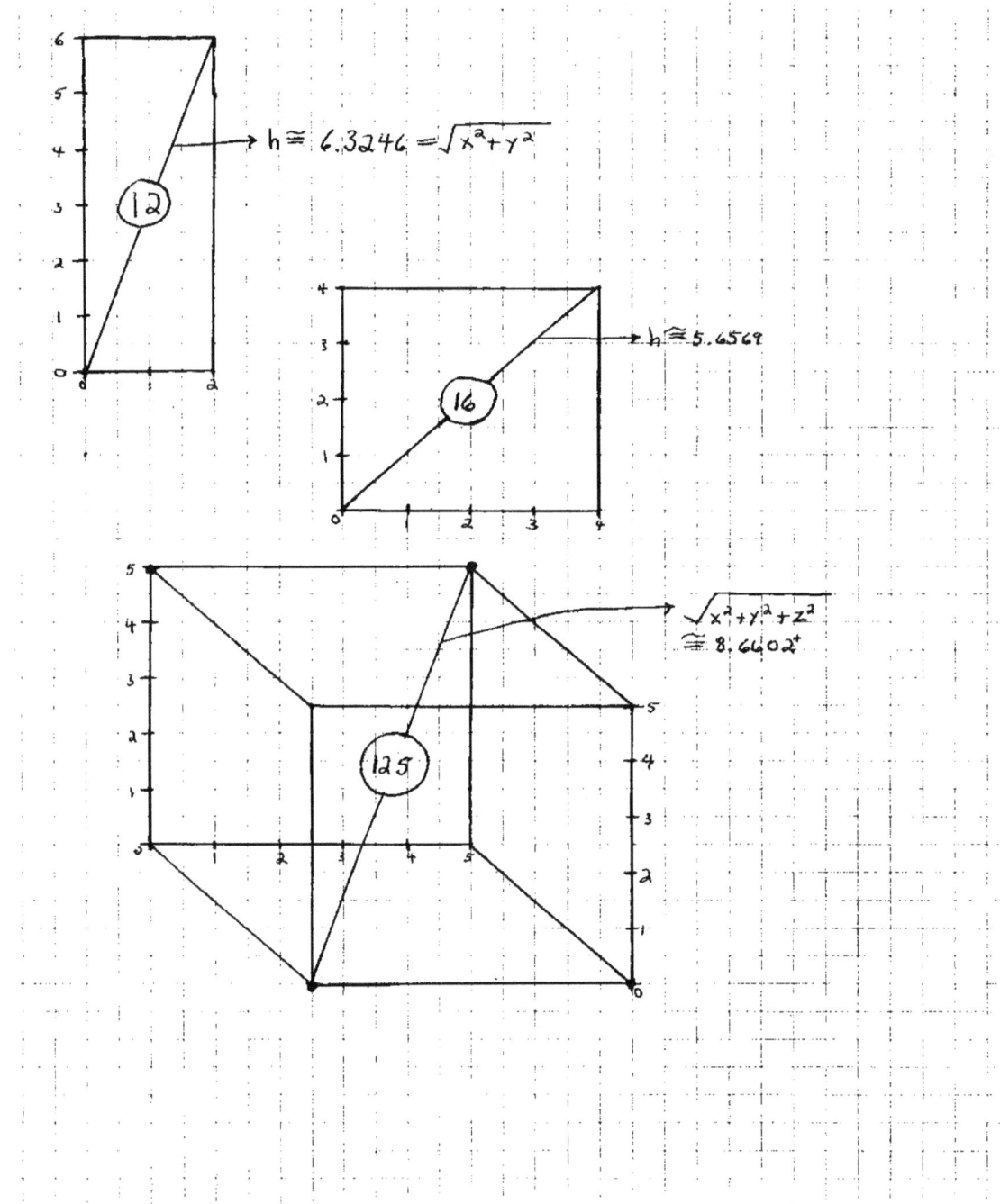

Graph One

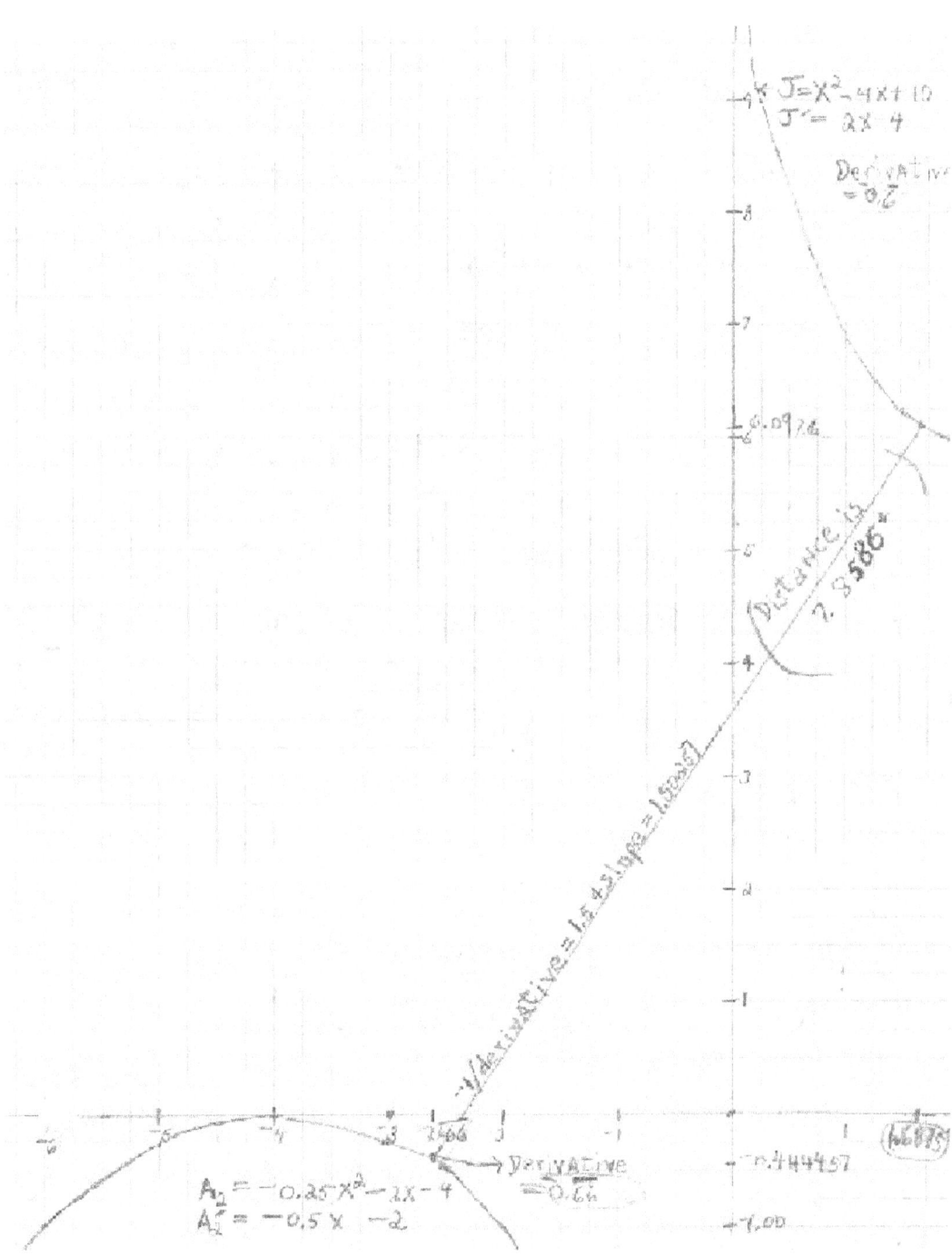

$J = X^2 - 4X + 10$
$J' = 2X - 4$

Derivative = 0

$\theta = 0.0924$

Distance is 7.3586"

-1/derivative = 1.5 x slope = 1.66667

Derivative = 0.6"

-0.444457

-7.00

$A_2 = -0.25 X^2 - 2X - 4$
$A_2' = -0.5 X - 2$

Graph Two

Here, the distance between these two functions is the length of a connecting line segment that connects where parabolas A_2 and J have derivative values of -0.666666. This line connects (-2.66666, -0.444457) to (1.6875, 6.0976). So, the slope is 1.50257 and this is close to the value of negative one divided by the derivative value of -0.6666666, or 1.50. The distance between these two parabolas is the length of the line segment that starts from either of the two above points. The length of this line segment is 7.8585.

$$(6.5421^2 + 4.3541^2)^{0.5} = 7.8586.$$

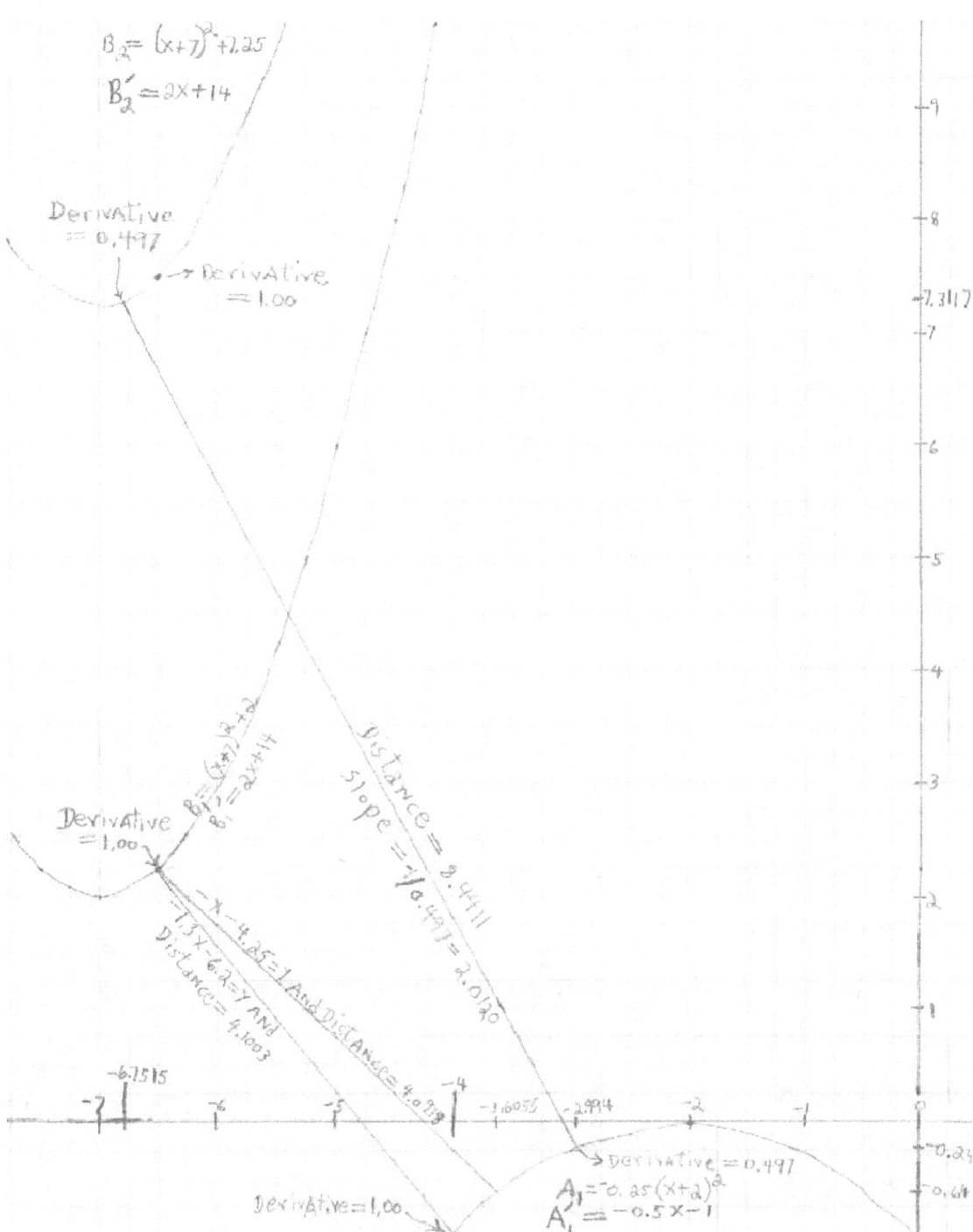

$B_2 = (x+7)^2 + 1.25$

$B_2' = 2x + 14$

Derivative = 0.497

→ Derivative = 1.00

9

8

7.3117

7

6

5

4

3

$B_1 = (x+7)^2 + 2$
$B_1' = 2x + 14$

Derivative = 1.00

Distance = 2.4411
Slope = 1/0.493 = 2.020

$-x - 4.25 = 1$ And Distance = 4.019

$-1.3x - 6.2 = y$ And Distance = 4.1003

2

1

-6.7515

-7   -6   -5   -3.6055   -2.994   -2   -1   0

-4

Derivative = 0.497

-0.25

-0.44

$A_1 = -0.25(x+2)^2$
$A_1' = -0.5x - 1$

Derivative = 1.00

Graph Three

256

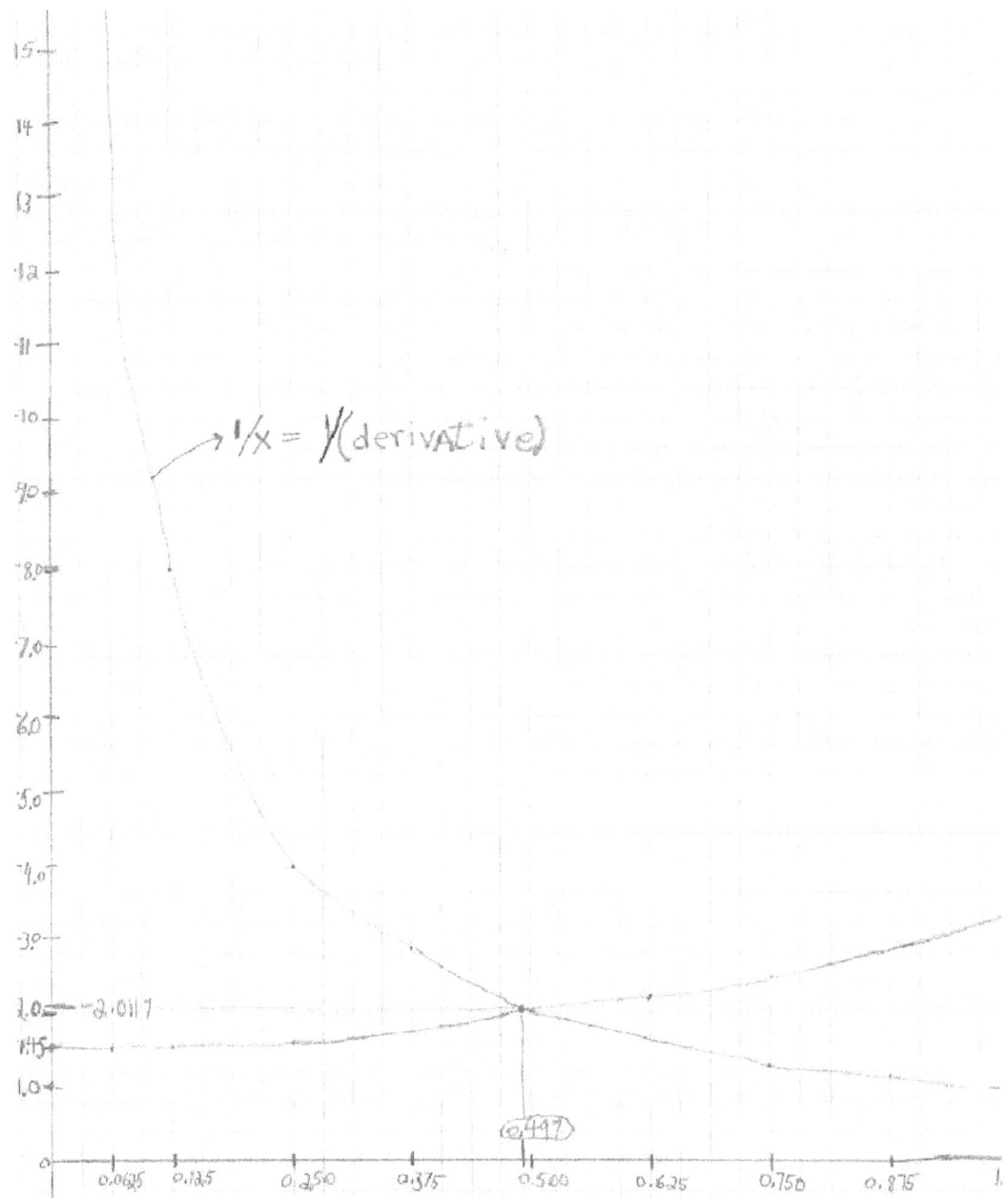

Graph Four

257

1.) Ten Connection Point

2.) Derivative Values of

3.) X & Y Values of Function $A_1 = -0.25(X+2)^2$ where the derivative is $A' = -0.5X - 1.00$

4.) X & Y Values of Function $B_2 = (X + 7)^2 + 2 = Y$ where the derivative is $B' = 2X + 14$.

5.) Deriving the Slope of the Connecting Line

6.) Slope of the Connecting Line

Nine Connection Points	Derivative Values of	X & Y Values of Function A_1 $= -0.25(X+2)^2$ where the derivative is $A' = -0.5X - 1.00$	X & Y Values of Function $B_2 = (X + 7)^2 + 2 = Y$ where the derivative is $B' = 2X + 14$.	Deriving the Slope of the Connecting Line	Slope of the Connecting Line
1	0.000	X = -2 Y = 0	X = -7.000 Y = 7.250	-7.25/5	-1.450
2	0.125	X= -2.25 Y= -0.015625	X= -6.9375 Y= 7.25390625	-7.26953125/4.6875	-1.5508333333
3	0.25	X =-2.5 Y = -0.0625	X = -6.875 Y = 7.265625	-7.328125/-4.6375	-1.675
4	0.375	X = -2.75 Y =-0.140625	X = -6.8125 Y = .28515625	-7.42578125/-4.0625	-1.827884615

Nine Connection Points	Derivative Values of	X & Y Values of Function A_1 = $-0.25(X+2)^2$ where the derivative is $A' = -0.5X-1.00$	X & Y Values of Function $B_2 = (X + 7)^2 + 2 = Y$ where the derivative is $B' = 2X + 14$.	Deriving the Slope of the Connecting Line	Slope of the Connecting Line
5	0.4971052	X = -2.994210526 Y =- 0.247113643	X = -6.7515 Y = 7.631175225	- 7.5586589/3.757289	-2.011732103 = about, the inverse of the derivative, 0.497.
6	0.625	X = -3.25 Y =-0.390625	X = -6.685 Y = 7.34765625	-773828125/3.435	-2.252774745
7	0.75	X = -3.5 Y =-0.5625	X = -6.625 Y = 7.390625	-7.953125/3.125	-2.545
8	0.8750	X =-3.75 Y = -0.765625	X = -6.5625 Y = 7.44140625	-8.20703125/2.8125	-2.91805556
9	1.000	X = -4.00, Y = -1.00	X = -6.5, Y = 7.5	-8.5/2.500	- 3.4

In the graph of $B_2 = (X + 7)^2 + 7.25$ a derivative value of 0.497 is at (-6.7515, 7.3117).

Also, $A_1 = -0.25 (X + 2)^2 = -2.994$

when $A_1' = -0.5X - 1.00 \quad = 0.4970$

and $Y = -0.247009$

and also $Y = -0.247.$

Connecting these two points with a line that has a slope that is near

$$-7.5587/3.7575 = -2.0116.$$

which is close to -1 divided by the common derivative of 0.497.

$$-1/0.497 \quad\quad = 2.0120.$$

Also, the answer 201178694 is about the same answer but it is an answer which has more decimal places.

Also, when finding the distance between two parabolas and connecting one point of each function, where they have equal derivative values, with a line that has a slope of -/+1, the common derivative value often has to be changed to (+/-1/ the common derivative value)$^{-1}$ because, often, the vertical distance between the two vertexes is smaller than the horizontal distance between the two vertexes.

Also, when finding the distance between Parabola B$_1$, $(x+7)^2$ + 2, and Parabola A, $-0.25(X+2)^2$, finding the distance between equal derivative values of one creates a distance of 4.100 but creating a line with a slope of negative one from where the Parabola B has a derivative value of one shows us that the distance between these two parabolas is 4.093. Here, we see the "strength" of where the parabola has a magnitude of one.

Of course when two parabolas are positioned in such a manner that their vertexes are vertically across from each other, here the slope of a line is one divided by zero or it is undefined. For two parabolas that are positioned in such a manner that their vertexes are horizontally across from each other, the slope of the line is zero but one divided by the derivative of one of either functions is one divided by zero. This is the case unless you pivot the graph so that the Y axis is positioned to be where the X axis usually is located. Now the vertex has a derivative of zero and one divided by this vertex is undefined.

Second, when you are working above where the more at being flat parabola's vertex is located, to find the distance between two parabolas you should connect the least flat parabolas' negative or positive one derivative value location to the other parabola by using a line with a slope of one or with a slope of negative one.

When working below the vertex of the more at being flat parabola, although the work can be as frustrating as a series of an abyss of conjugated asymptotes, the two parabolas have a distance between them that can be found by

drawing a square that has opposite sides that touch each parabola and that are connected by a hypotenuse line that separates this square figure into two triangles.   This is illustrated with the parabolas $A_3 = -(\{X^2 - 4X + 4\}/4)$ where the derivative is $0.5 X - 1$ and $D = (X-8)^2 - 7$, where the derivative is $2 X - 16$, and when x is 7.5 the derivative is negative one.   The distance between them is found by connecting where the first of them has a derivative value of -2.375 and the second of them has a derivative value of -2.368. Here, the square, for lack of a better word, has coordinates of (6.7500, -5.6500), (6.8166, -5.6500),

& (6.8166, -5.5833), (6.7500, -5.6500).

The Square

263

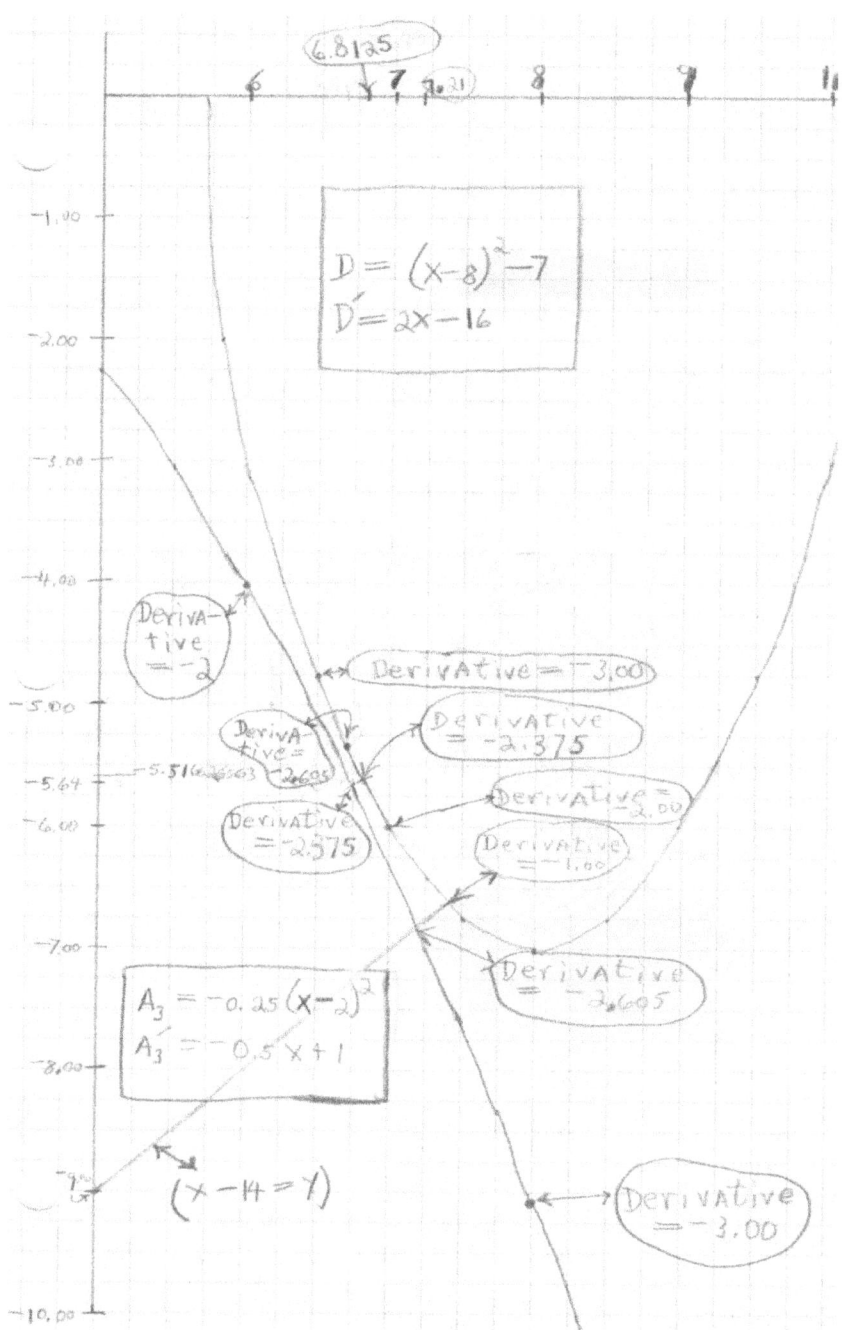

6.8125

6   7 7.31   8   9   11

$D = (x-8)^2 - 7$
$D' = 2x - 16$

−1.00

−2.00

−3.00

−4.00

Deriva-
tive
= −2

Derivative = −3.00

−5.00

DerivA-
tive =
−2.605

Derivative
= −2.375

−5.64   5.516...63

Derivative =
−2.00

−6.00

Derivative
= −2.375

Derivative
= −1.00

−7.00

Derivative
= −2.605

$A_3 = -0.25(x-2)^2$
$A_3' = -0.5x + 1$

−8.00

−9.00 $(x - 14 = Y)$

Derivative
= −3.00

−10.00

<u>Graph Five A</u>

264

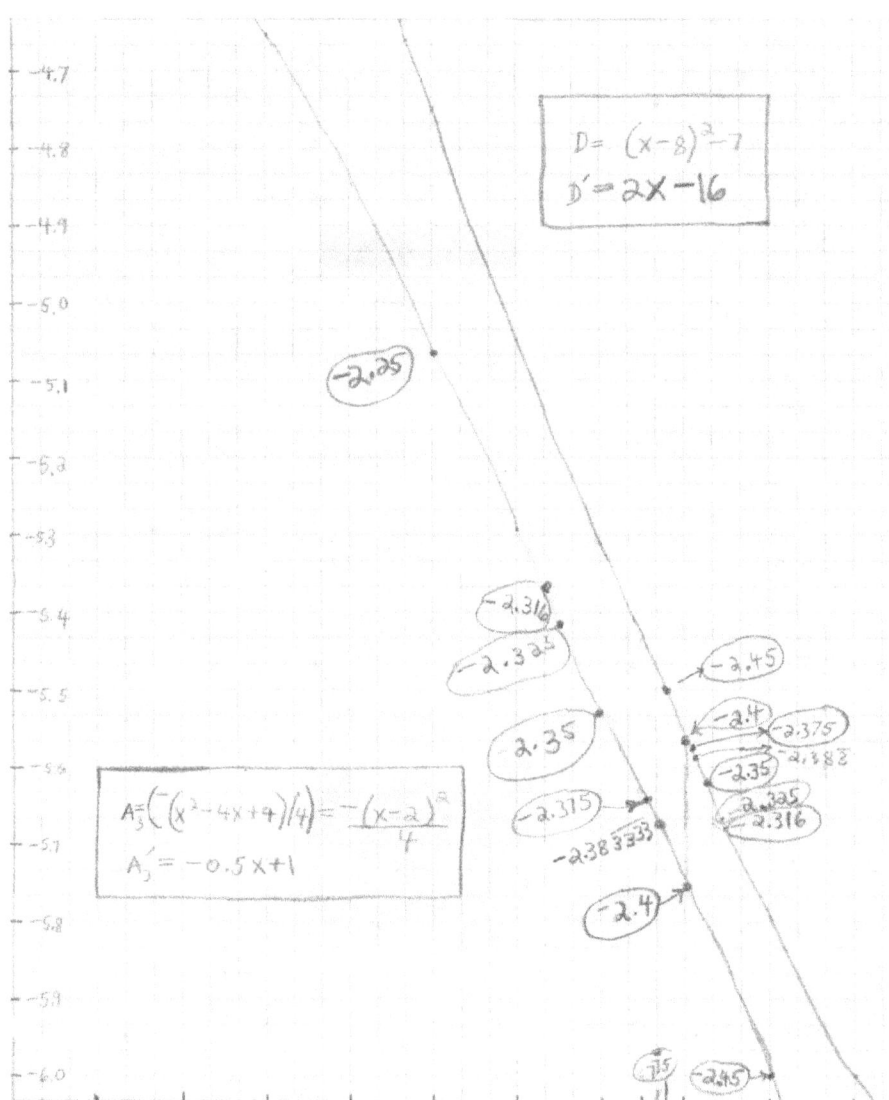

Graph Five B

265

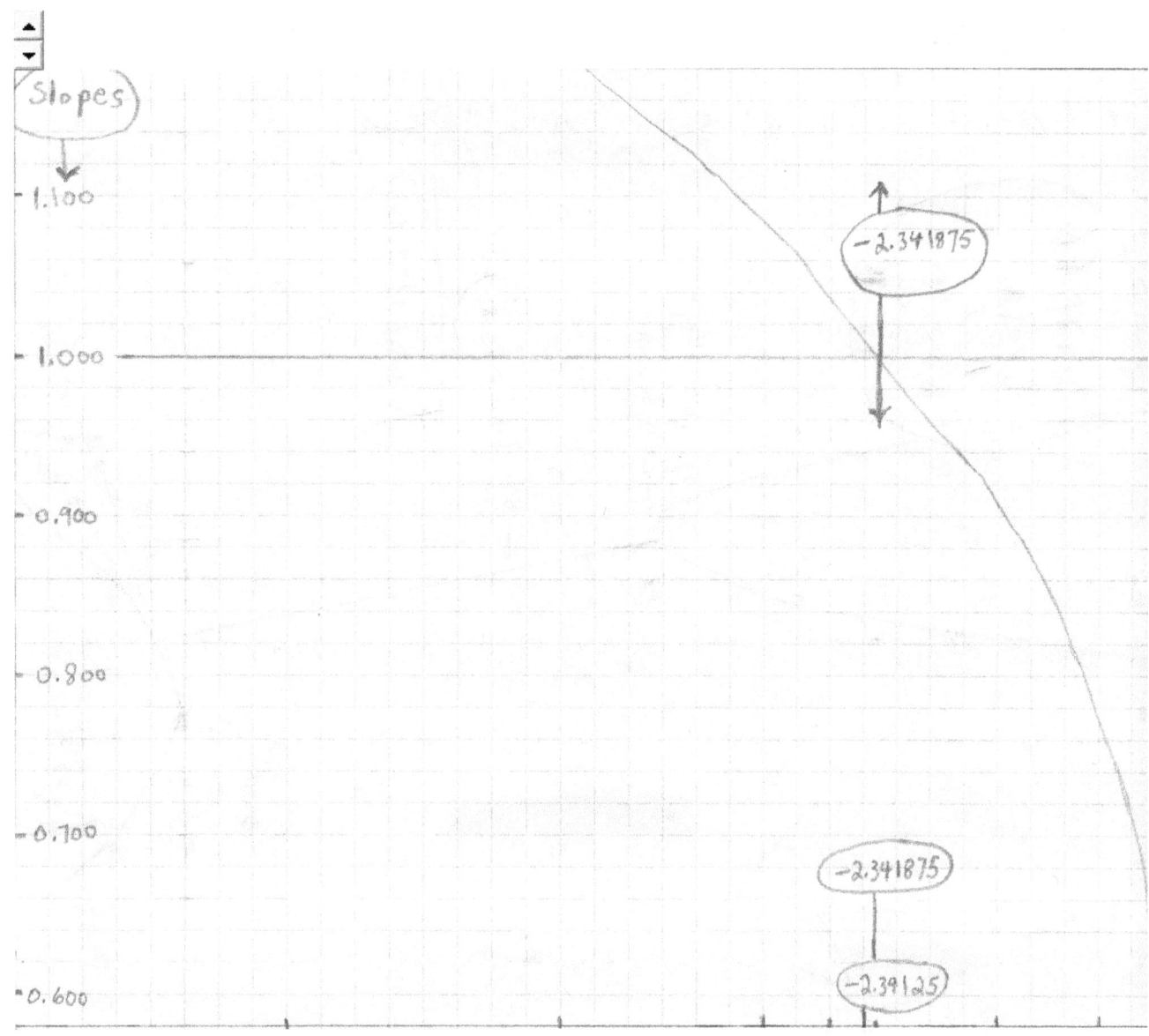

Graph Six

In the following example of **how to find the distance between a vertical and a horizontal parabola** we will observe both parabola curves F and G.

The formula of F is $F = -(X^2 + 4)/2$.

It can be expressed as $F = (-2Y - 4)^{1/2}$.

The derivative of F is one at (1, -2.5), not a location that we should use to find the distance between these two parabolas, F and G.

The formulas of G is $G = 0.125Y^2 - 0.125Y + 2$

Where the derivative of G is one is irrelevant.

When function F has a derivative value of 0.6455, at (1.5495, 3.20), and when function G has a derivative value of 0.625, at (2.75, -2.00), a line with a slope of 1.00 connects them. The length of this line segment of X - 4.75 = Y is 1.6974.

Another example of the distance between two non-intersecting parabolas, where one is vertical and the other is horizontal, is the example of parabolas

$-X^2/5 -2X/5 -7/5 = H$ that has a derivative of $(0.4X -0.4) = H'$ and

$-(-X-1)^{0.5}$ or $(-Y^2 -1) = I$ and that has a derivative of $0.5(-X -1 = I)^{-0.5}$. Here, the distance between them is found by connecting similar derivative values with a hypotenuse line of a rectangular figure.

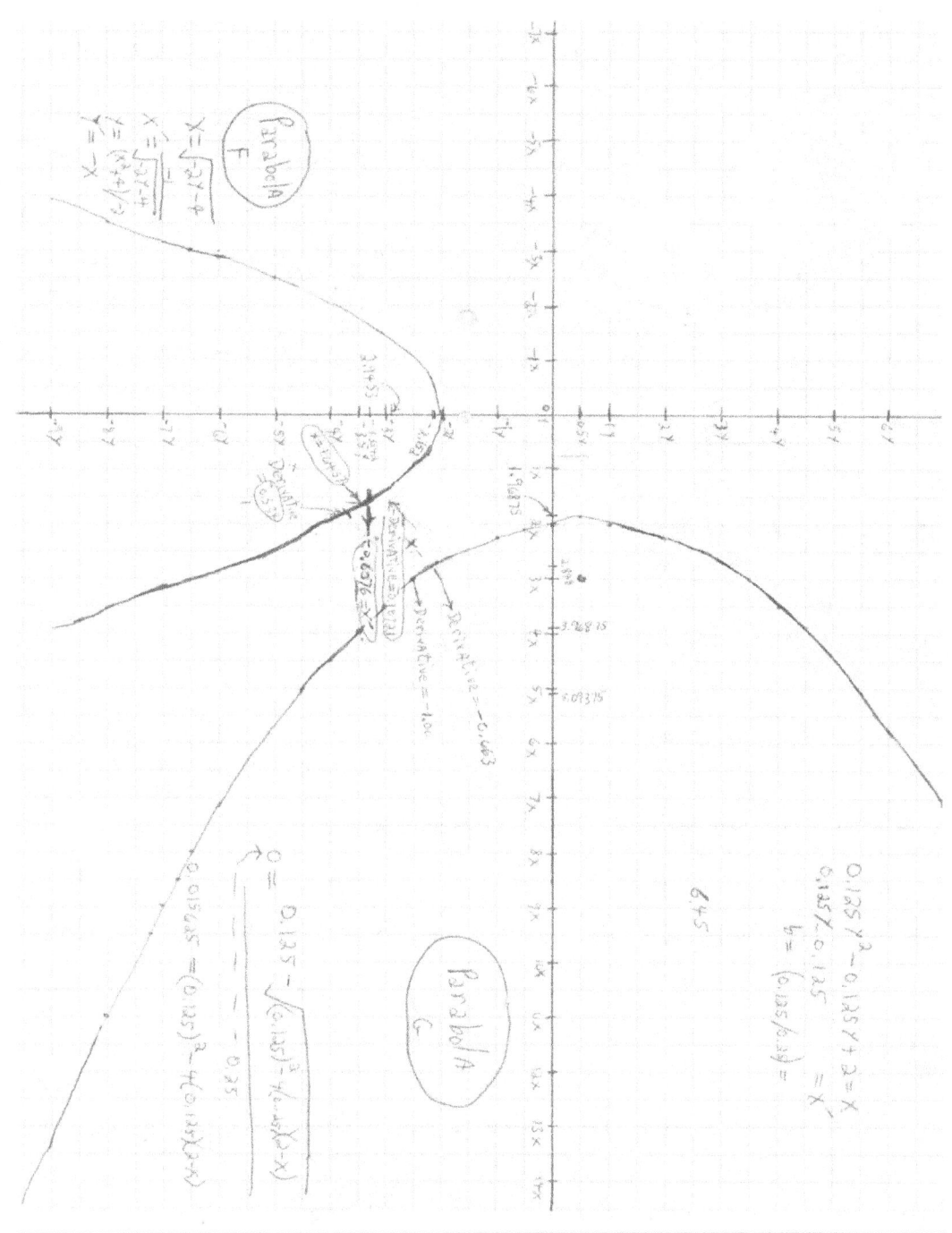

Graph Seven

Graph Eight

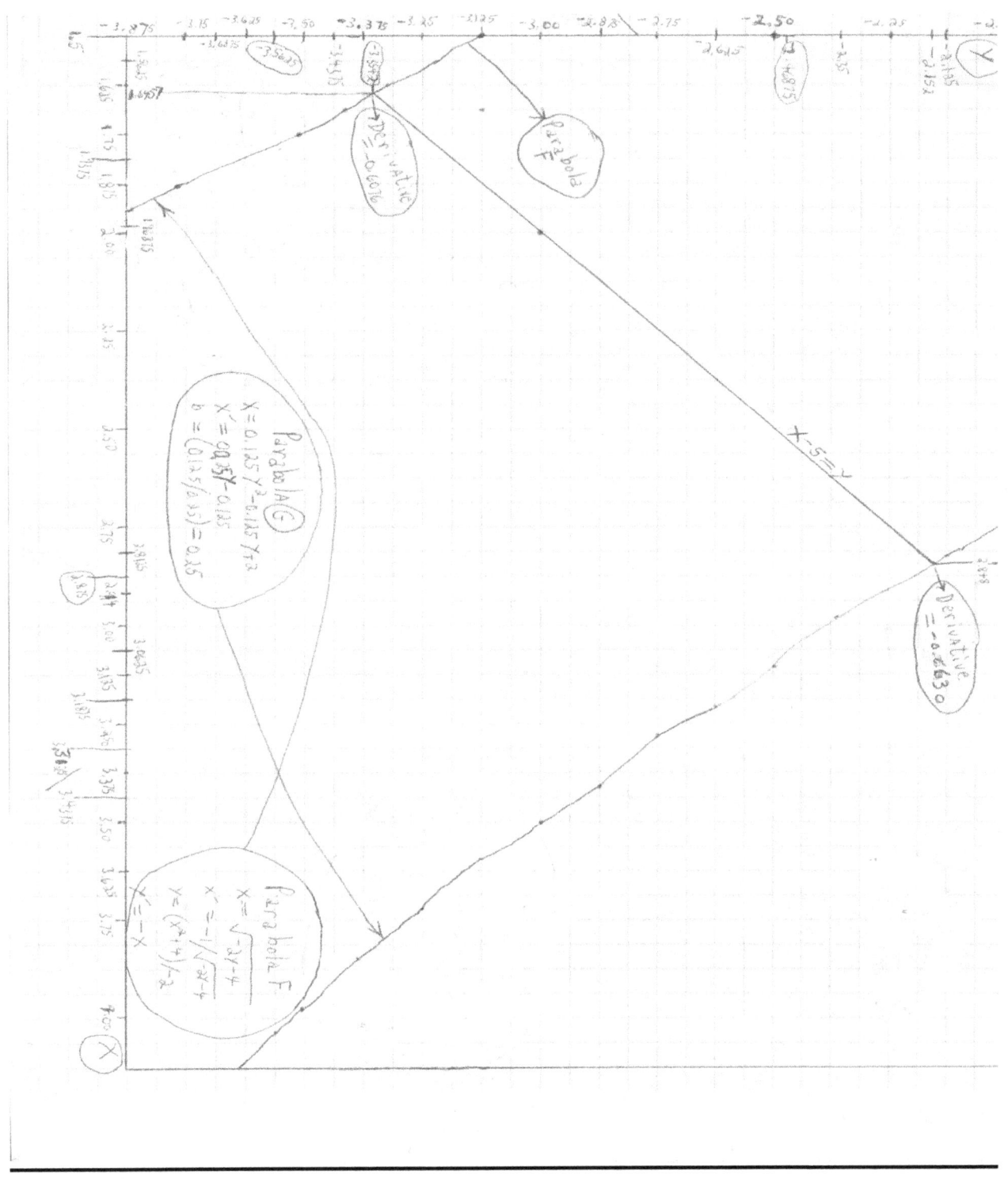

Graph Nine

Graph Nine A

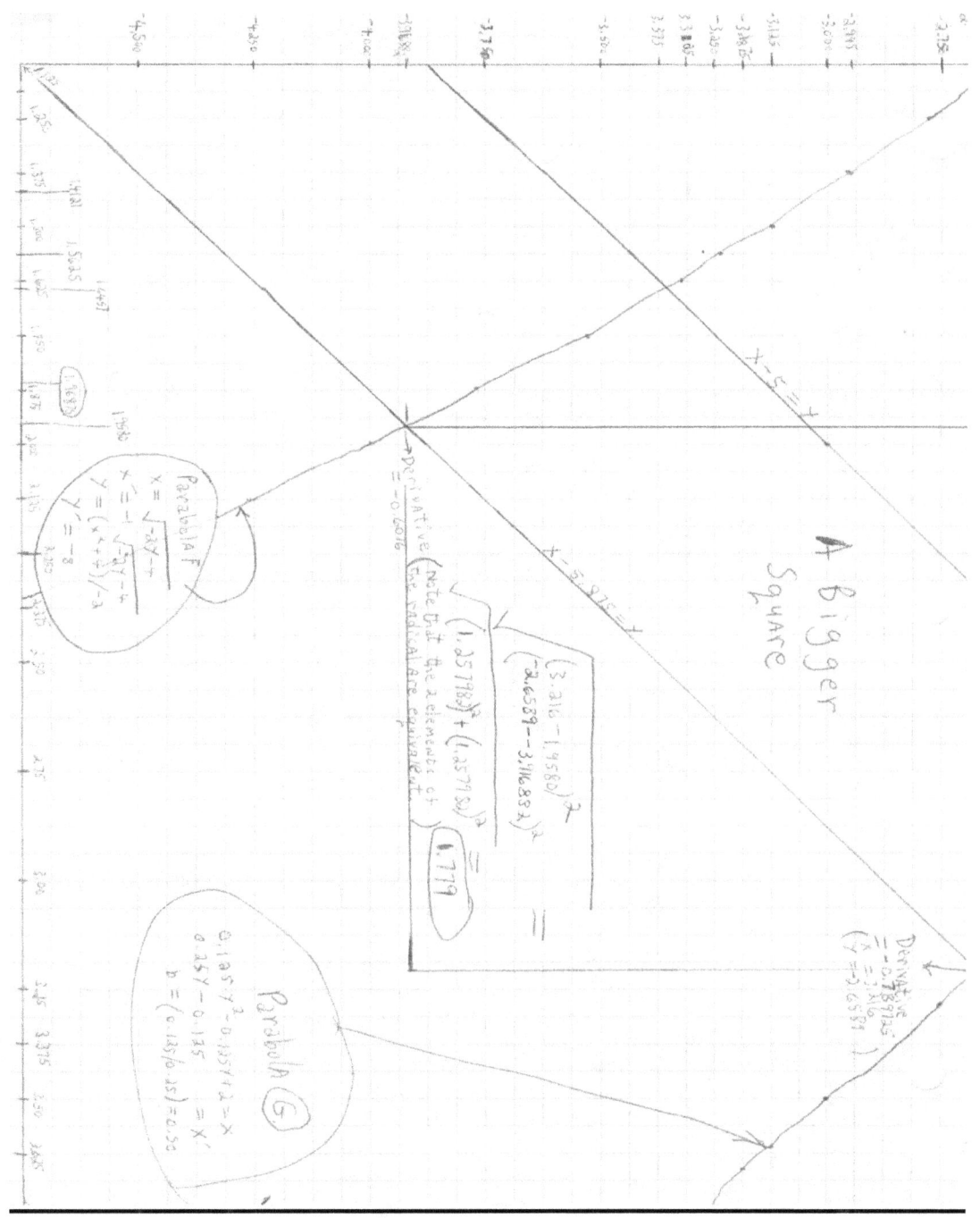

Graph Nine B

272

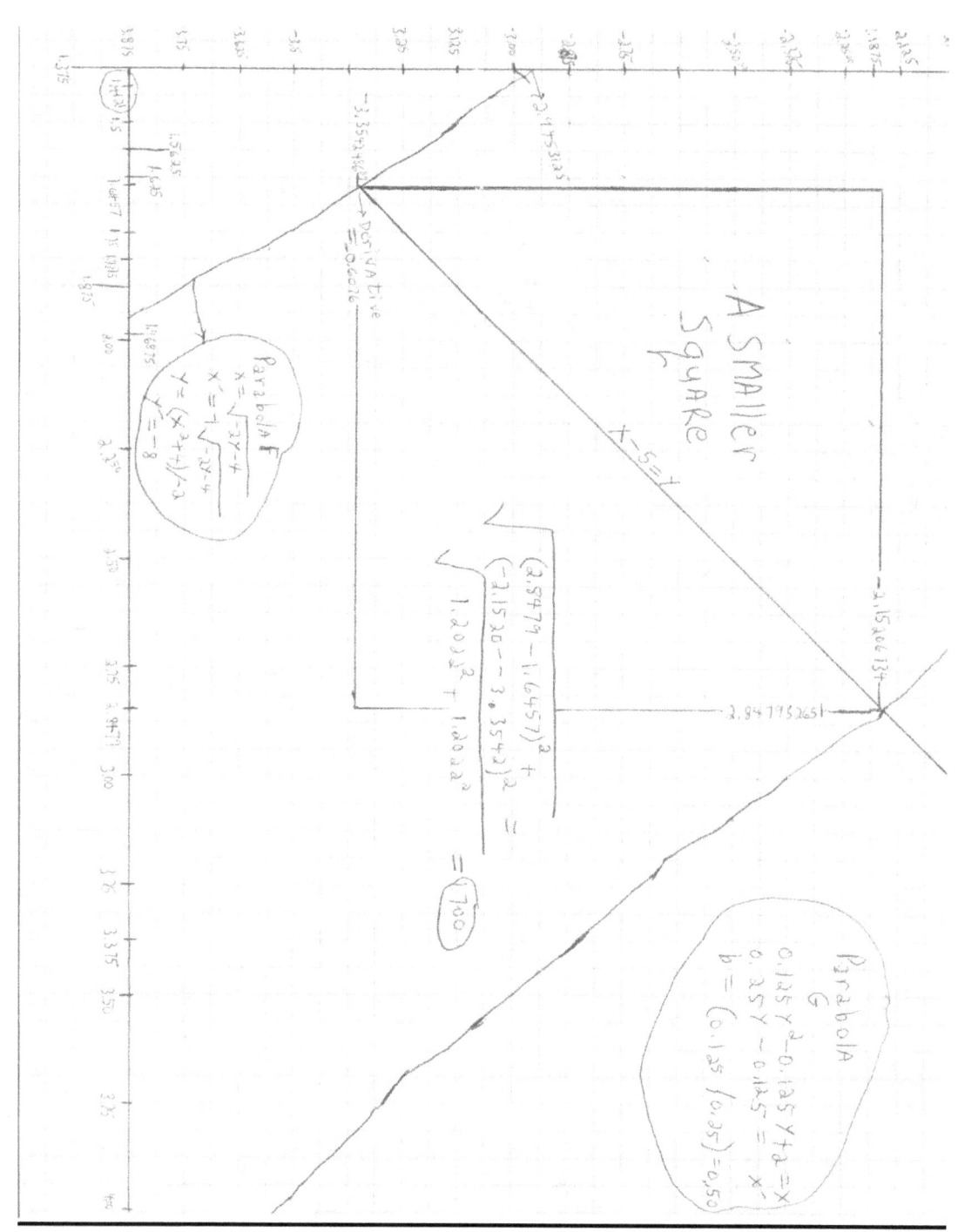

Graph Ten

273

Some results of the connection between Parabolas H & I

The vertical line that starts from the X axis in the horizontal parabola that is named parabola H and turns forty-five degrees at the bottom of this parabola before it connects to the vertical parabola that is named parabola I.	Derivative Values	Derivative Values	(Derivative differences) (Sum of derivatives)/2
	The Top Parabola	The Bottom Parabola	
-0.750	Infinity	-0.1000	0.2000
-1.500	0.3535	0.2000	0.5547
-2.125	0.5303	0.4500	0.1638
-2.375	0.5863	0.5500	0.0638
-2.500	0.6123	0.6000	0.0204
-2.7500	0.6614	1.2123	0.5881

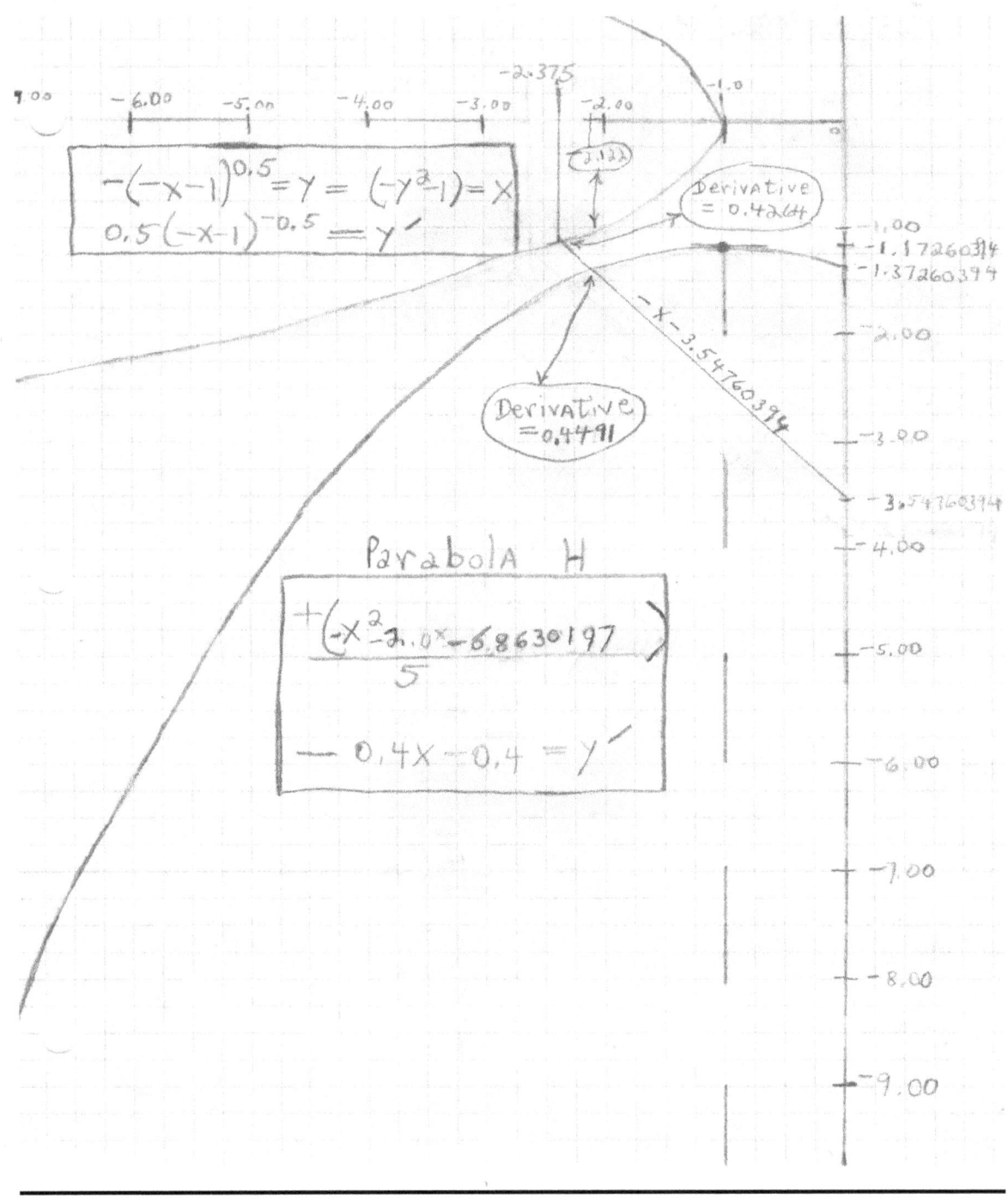

$$-(-x-1)^{0.5} = Y = (-Y^2-1) = X$$
$$0.5(-x-1)^{-0.5} = Y'$$

−2.375

−1.0

(2.122)

Derivative = 0.4264

−1.00
−1.17260394
−1.37260394

−X−3.54760394

Derivative = 0.4491

−3.54760394

Parabola H

$$\frac{+(-X^2-7.0^x-6.8630197)}{5}$$

$$-0.4X-0.4 = Y'$$

One can slide parabola H along a forty-five degree line to where it touches parabola I.

Much more work is needed, still, for me to understand how the distance between two non-intersecting horizontal and vertical parabolas exist.

For now a few things seem possible. The distance between where Parabola I has a derivative value of 0.4264, which is adjusted to become 0.5 after it is multiplied by Parabola H's Y vertex value, can be connected to parabola H's derivative value of 0.50. Here, the distance between these two points is smaller than that which is shown in the above graph. Moreover, it seems that there are two places that can be connected by a line with a slope of negative one. This line will connect a location of parabola I that, after being multiplied by 1.17260394, equals the derivative position of parabola H of about 0.475. This line might be found to be the real distance between the two parabolas.

PARABOLAS FOR DUMMIES

Let's find the Focus and the Directrix

Directrix = -b/4a + C -1/4a according to most literature of parabolas.

To find the focus let's move from where the focus is expected to be horizontally. We will reach a place on the parabola that has a derivative value of one or negative one. So, to find the focus, simply find out where on the parabola is there a derivative value of either one or negative one and use a coordinate of that value, either an X or a Y value, that intersects the line of symmetry. Where the intersection of either this X or Y crosses the axis of symmetry is where the focus is.

Again, by setting the derivative of the function to equal the number one,   2X + 2 = 1.00,

we find that X equals negative one-half. Placing negative one-half into the formula gives

$$Y = (-0.5)^2 + 2(-0.5) = -8.75.$$

We found the vertex to be at (-1, -9) and so the focus, which is on the line of symmetry, is at

(-1, -8.75).

To find the directrix we only have to draw a line from where the parabola has a derivative of one or negative one to the axis of symmetry and create a line there that is perpendicular to the axis of symmetry.

THIS LINE HAS A SLOPE OF EITHER ONE OR NEGATIVE ONE.

QUESTIONS

What happens when he axis of symmetry is not a horizontal or vertical line and can the axis of symmetry be a line such as the X = Y line?

Can this line of symmetry not be a linear line and can it oscillate?

Here is some more information about parabolas.

Suppose there is a vertical parabola near a point that is below where there is a straight line that has a slope of one, or 45 degrees, and that extends out from where the parabola has a derivative value of 1.00. How can we find the distance between it and the parabola?

The answer can be found by extending a line with a 225 degree angle from the point to the parabola. Let the parabola be equal to

$$-1.25X^2 + 2X + 3.$$

Here, $-b/2a$(which is the formula to find the access of symmetry) is equal to $-2/-2.5$ and this corresponds to a point of (0.80, 3.80).

What if the point is above where there is a straight line that has a slope of one, or 45 degrees, and that extends out from where the parabola, parabola B, has a derivative of -1.00. This point has also a "Y" value that is less than the "Y" value of the parabola's Directrix and the point has a "Y" value that is less than the " value of the parabola's vertex "Y" value. This point is (1.25, 3.75).

To find the distance between the point and parabola B, where parabola B is ($-1.25X + 2X^2 + 3$), we can create an imaginary parabola, parabola C, that has the same distance from the point as does the original parabola, parabola, but in an opposite direction. By seeing that the distance between parabola B's vertex and the point (1.25, 3.75) is (0.45, -0.05, the opposite parabola is constructed with a vertex of (1.7, 3.7). To make the new parabola's access of symmetry to become the X line of 1.7, -b/2a has to equal 1.7 so –b has to be 4.25. Here, -b/2a equals 4.25/2.5 and the result is 1.7. Rather than create large changes we'll keep the "A" term as it was. The "C" term has to be, then, 7.3125 to make the "Y" value of the vertex to be 3.7.

$$\text{Parabola B} = -1.25 + 2.00X^2 + 3.0000$$

$$\text{Parabola C} = 1.25X^2 - 4.25X + 7.3125$$

Let's find the distance between these two parabolas, parabola B and parabola C.

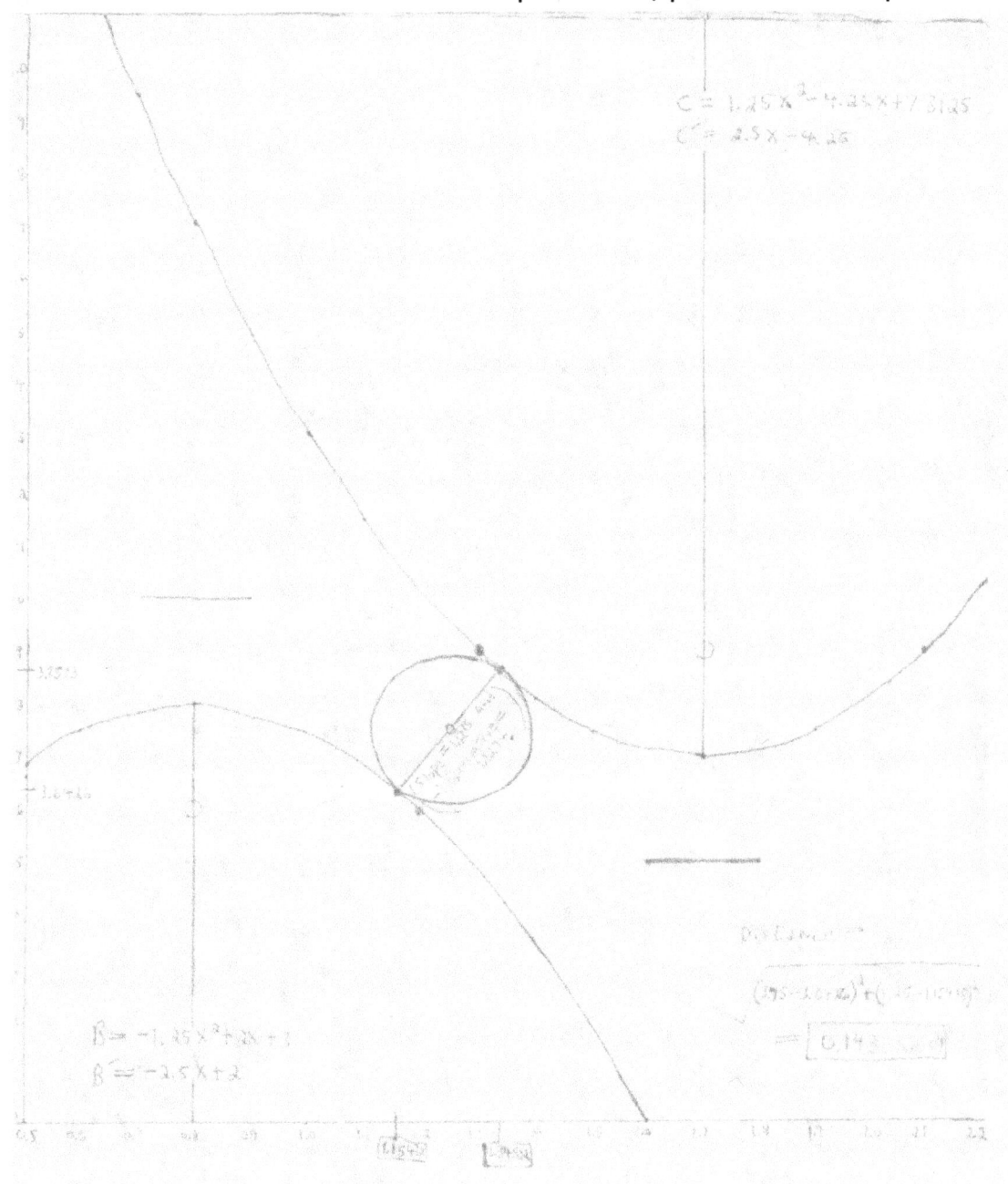

Graph 12

Graph12

Parabola B is $-1.25X^2 + 2X + 3$

Parabola C is $1.25 X^2 - 4.25 X + 7.3125$

	Parabola B =				Parabola C =		
	$-1.25X^2 + 2X + 3$					$1.25X^2 - 4.25X + 7.3125$		
	Parabola B' =					Parabola C' =		
	$-2.50X + 2$					$2.50X^2 - 4.25$		
Derivatives	X	Y	Slope	-1/Derivative	Derivatives	X	Y	
-0.8860	1.1544	3.6430	1.1192	1.1286	-0.886	1.3456	3.8569	
-0.8865	1.1546	3.6428	1.1231	1.1280	-0886	1.3454	3.8571	
-0.8866	1.1546	3.6427	1.1238	1.1279	-0.886	1.3454	3.8571	
-0.8869	1.1547	3.6426	1.1268	1.1275	-0.886	1.3452	3.8573	
-0.8870	1.1548	3.6426	1.1279	1.1273	-0.887	1.3452	3.8573	
-0.88695	1.1547	3.6426	1.1275	1/1274	-88695	1.3452	3.8573	

Parabola B is $-1.25X^2 + 2X + 3$
Parabola C is $1.25 X^2 - 4.25 X + 7.3125$

Since the two shaded boxes have almost the same value we can see that

the slope of the line that connects derivative values of -0.8870 is equal to the

inverse of the common derivative values of the two parabolas. By definition, this

line is the shortest line between the two parabolas.

This line also passes through the point that is located at (1.25, 3.75).

The distance between the point and Parabola B is

$$[(3.75 - 3.6426)^2 + (1.2500 - 1.15478)^2]^{0.5} = 0.143532604$$

The distance between the point and Parabola C is

$$[(3.85 - 3.7500)^2 + (1.3452 - 1.25000)^2]^{0.5} = 0.143489123$$

What is the distance between the parabola and a point
if the point is above the directrix and is located at (1.075, 4.25). Here,
we find the distance by constructing an imaginary parabola, Parabola
B_2, that is an equal distance from the point but in an opposite
direction as is Parabola B from the point and it is the same as
Parabola B_1 but it is facing up instead of facing down.

Here the parabola that faces down is
$$B_1 = -1.25 X + 2X^2 + 3.$$
The parabola that faces up is
$$B_2 = 1.25 X - 3.375X + 6.978125.$$
The coefficients of the formula were found by keeping the first term,
the a term, as it was in magnitude but changing it's direction from
negative to positive. The center of the parabola can be found by
using the formula $-b/2a$ = an access of symmetry that is an equal
distance, but in an opposite direction, from the point as is the access
of symmetry of parabola B_1 from the point.

See Graph 13

To find the formula of B_2 the first term's sign was changed, the
access of symmetry had to be an equal distance, but opposite in
direction, from the point as is parabola B1 from the point and it had to
be constructed from the term of $-b/2a$. So, the B_2 parabola must
have an access of symmetry of 1.35 and the "a" term is -1.25. So, -
b/2a has to equal 1.35.

So, with $-b/2.5$ equaled to 1.35 we can see that b equals -3.375.

To make the vertex of Parabola B_2 to have an equal distance, but in an opposite direction,
from the point as does parabola B_1, 6.978125 has to be the "c" term. Parabola B_1 is $1.25X^2 -$
$3.275X + 6.978125$ and its derivative is $2.50X - 3.375$. The point becomes the center of the circle
that intersects both parabolas where they have the same derivative value of -0.316.

The top parabola is $B_2 = 0.125X^2 - 3.375X + 6.978125$
and $B_2 = 2.5X - 3.375.$

285

| $B_1 = -1.25 X^2 + 2X +3.$ | | | | $B_2 = 1.25 X^2 -3.375X + 6.978125.$ | | |
| $B_1' = -2.50X^2 +2.$ | | | | $B_2' = 2.50 X -3.375$ | | |

Derivative	X	Y	Slope	-/Derivative	X	Y
-0.315	0.926	3.78053	3.138154	3.174603	1.2252	4.719468
-0.316	0.9264	3.78002	3.162659	3.164556	1.2236	4.719971
-0.317	0.9268	3.7799	3.17205	3.154574	1.2232	4.720098

We can see in the table that the slope of the line that connects the

where both parabolas have a derivative value of -0.316 is close to the

value of -1/ -0.316.

36

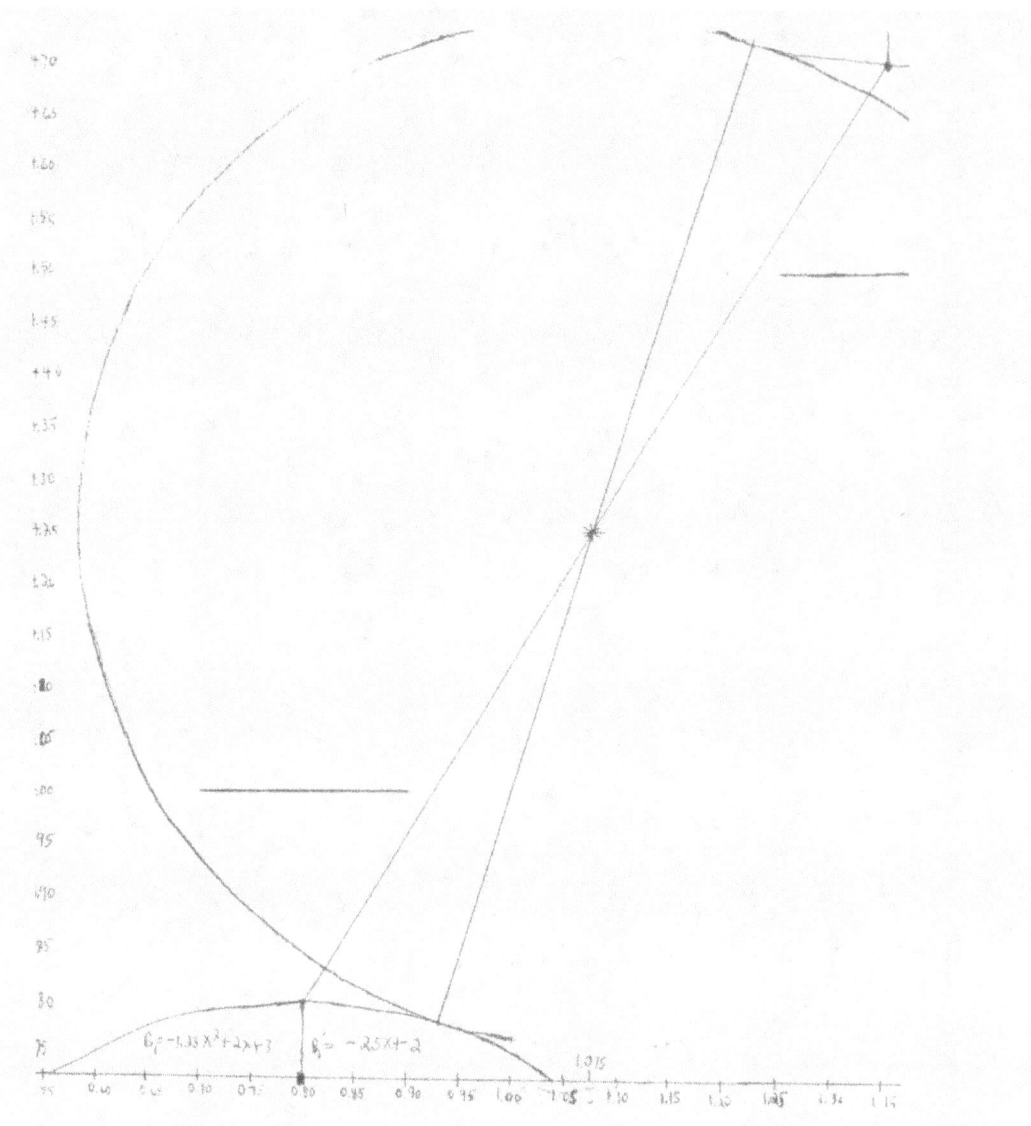

$g_1 = 1.33x^2 + 2x + 3$ $g_2' = -2.5x + 2$

1.015

Graph 13

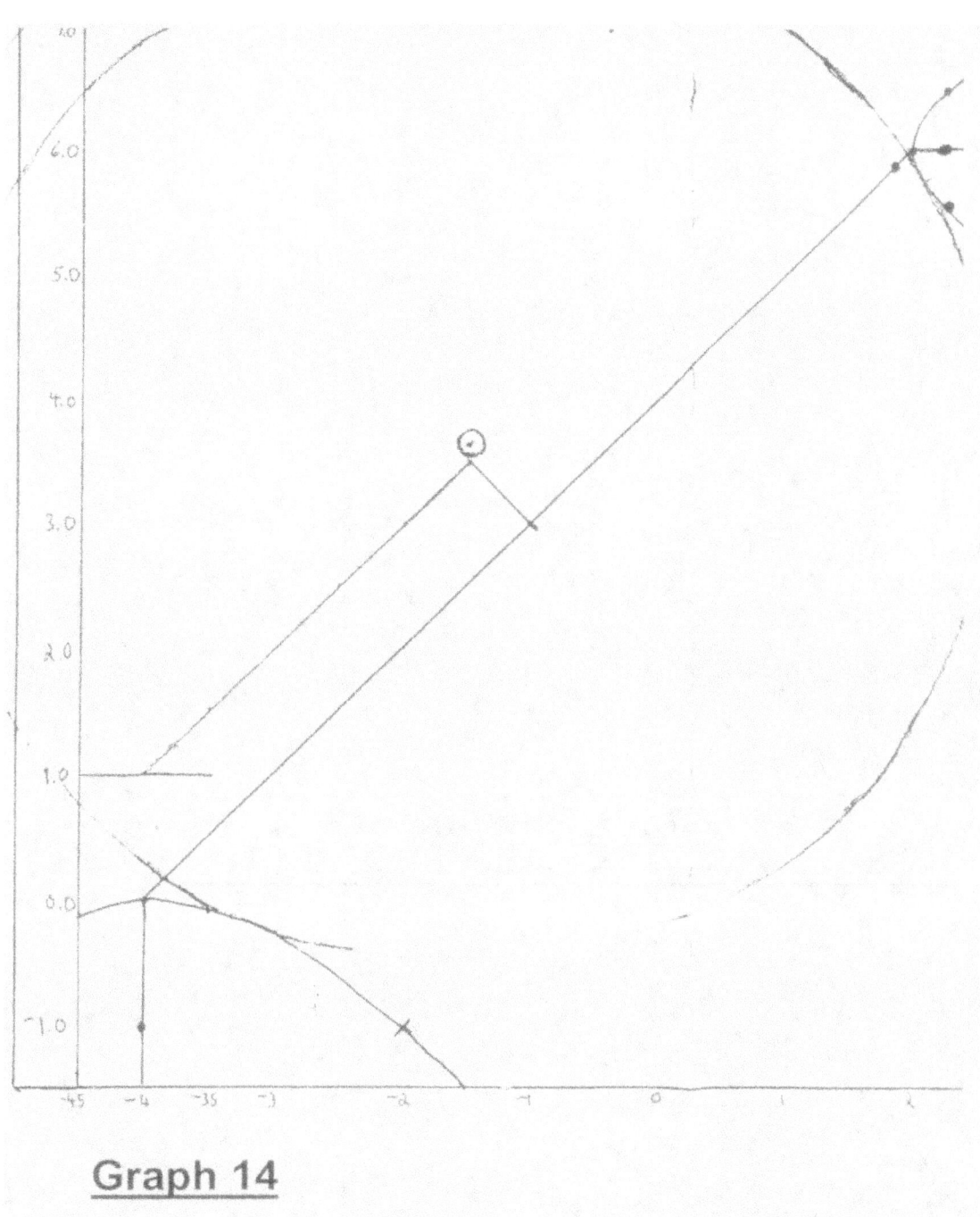

Graph 14

288

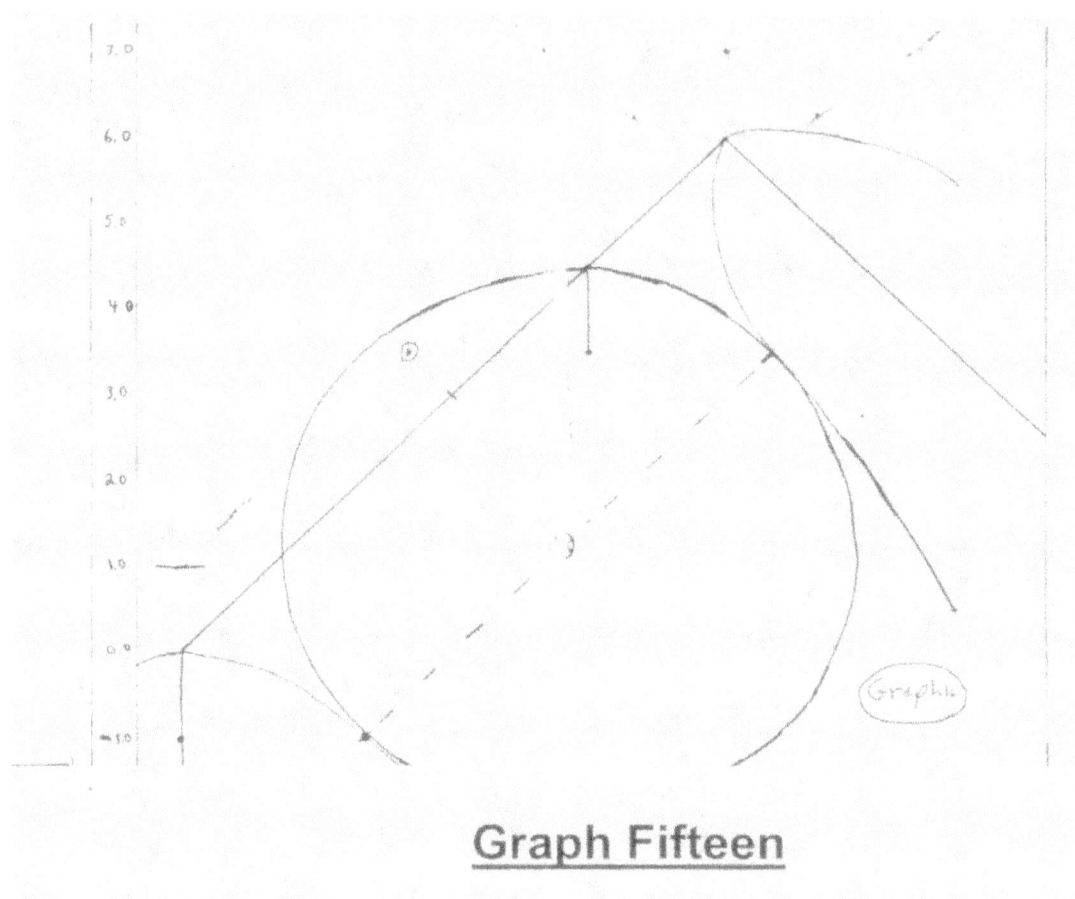

Graph Fifteen

Graphs fourteen and fifteen show us what happens when we are dealing with the distance between two non-intersecting parabolas that both are not vertical, horizontal.

To find the distance between two non-intersecting parabolas that both are not vertical or horizontal, orientate the entire graph so that the least flat parabola of the two, if they both are not equally flat, becomes a vertical parabola. Here, should the vertical parabola intersect with a straight forty-five degree or one hundred and thirty-five degree line that also intersects with the other parabola then the distance between the two functions can be found by connecting them with a straight line that has a slope of one or negative one and that connects to where both parabolas have the same derivative values. This is similar to what was done in graph six.

Otherwise, one place of intersection is where the least flat parabola has a derivative value of one or negative one and the line that intersects with this parabola has a slope of one or negative one.

Also, in both situations, a circle can be drawn between the parabolas. Where this circle touches, once, each of the two parabolas is where the distance between the two parabolas is minimized and the two derivatives of the points of intersection with the circle are the same. Note: that this line that connects the two parabolas does not always pass through the center of the circle. The radius of the circle is the distance between the circle's center and either of the two parabolas

A Relationship between
Circles, Squares, Spheres and Cubes *

(Please do not just be impressed with the numbers and words and assume that I must be correct but be skeptical and read everything.)

What are some relationships between these elements? To show some let us observe a circle with a radius of three and, thus, an area of 28.27433388 with, going further, a surface area of 18.84955592 and compare it with a square that has the same size perimeter as the circle has as its circumference.

Circle's Area $= \pi r^2 =$ 28.27433388
Circle Circumference $= 2\pi r =$ 18.84955592
Circle Radius $= 3$, given
Square's Area $=$ Multiplication of one side X by one side Y.

There is a square with each side of it having one quarter of the circle's circumference as its length. Here, 18.84955592/4 = 4.71238898. So, the area of the square is 4.71238898^2 or 22.2066099.

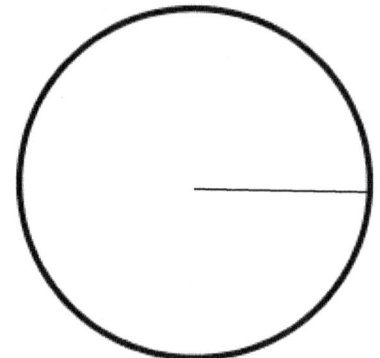 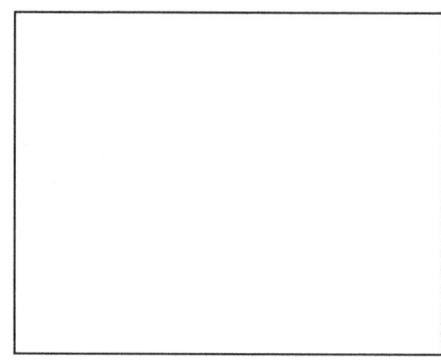

The radius is three feet long and it is the line that is inside of the circle.

The area of the circle divided by the area of the square is

$\underline{28.27433388}$ = 1.273239545. This is the First Very Important number, 4/π.
22.2066099

Also the diameter of the circle is 1.273239545 longer than any side of the square.

Here, 1.273139545 is always found in this relationship when a circle and a square have a similar size circumference and perimeter.

When the radius is one, area is pi, circumference is two pi times the radius, r, and a side of the corresponding square is (pi)(r)/2. Volume of the square is [(pi)(r)/2]^2. So the ratio of a circle's circumference to a square's area is

6.283185307/ [(pi=π 0r 3.1415....)(r)/2]^2

Here is another explanation of this.

$2\pi r/(2\pi r/4)^2]$ = $2\pi r/(0.5\pi r)^2$ = $(2)(4)/(\pi r)$ = $(4/\pi)(2/R)$.

This TIMES ONE-HALF THE CIRCLE'S RADIUS = 1.273239545.

A Relationship between
Circles, Squares, Spheres and Cubes *

Next let us observe a sphere and a cube where a similar relationship exists.

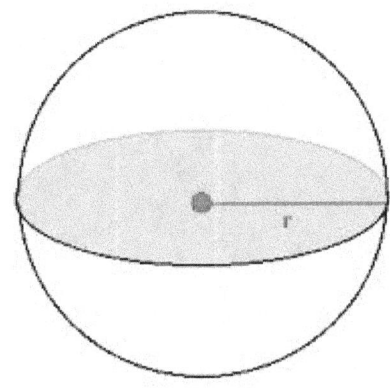

-

Volume = $\frac{4\pi r^3}{3}$ = 113.0973355

Note that the derivative of the volume is the surface area.

Surface Area = $4\pi r^2$ = 113.0973355

Radius = 3

Since this surface area is the above, it divided by the six sides of the cube is 18.84955592 and the square route of this is the length of any side of the cube, 4.3415607527. Going further, a length side of the cube taken to the third power, or cubed, is the area of the cube, 81.83737384. Dividing it into the volume of the sphere shows the relationship between the sphere and the cube as being 1.381976598,

A Second Very Important Number.

Next, we can find the relationship between the two Very Important Numbers.

1.381976598   _____ = 1.085487136

                            A Third Important Number

1.273239545

These three numbers will be found whenever circles, squares, spheres, and cubes are observed to have similar circumferences, perimeters, and surface areas.

A Relationship between
Circles, Squares, Spheres and Cubes *

When a square and a circle have the same size of perimeter and circumference, respectively, the area of the circle will be 1.273239545 times the area of the square.

$(\pi)r^2$ = Area of a circle.

$2(\pi)r$ = the circumference of a circle.

The letter r is the radius of a circle. When it, the radius, has a size of one the area of the circle is pi or 3.141592654 and the circumference is 6.283185307. When a square has this value as its perimeter it has an area value of 2.467401101, because with four equal sides sharing the value of 6.283185307 creates 1.570796327 for each side. Multiplying two non-opposite sides of the square together gives the square's area of 2.467401101. The relationship of these two areas is

3.141592654/2.467401101

And this shows that the circle has 1.273239545 more area when it has a circumference equal in size to the square's perimeter.

This relationship continues with cubes and spheres.

$$\frac{4(\pi)r^3}{3} = \text{the volume of a sphere}$$

The derivative of that is $4(\pi)r^2$ which equals the surface of a sphere.

We can compare the circumference of a circle with the perimeter of a square but if we expand to a sphere we must look at the surface of the sphere because it has no circumference and we can compare it to a cube, instead of a square.

Let us start with a surface of a sphere that is 24. Here r = 1.909859317and the area, A, is 29.18050088.

In this example, let the surface of the cube be 24. Here, each surface side has a value of four, 4, and the length of each side of the cube is two, 2. So, the volume is 2^3 or 8.

Next we will see some relationships. The relationship of the volume of the sphere to the volume of the cube is

29.18050088/8

Is there a relationship of this fraction with the very important number above? Yes, where the important number is 1.273239545.

(29.18050088/8)(4/9) = 1.273239545^2

Note that the above number that is being squared is the same enclosed number above.

Also, (29.18050088/8) (X) = 1.273239545. Here, X has a value of 0.34906585 which is about π (pi) divided by nine, 9.

Again, (29.18050088/8)(π/9) = 1.273239544, which is close to the original enclosed
number above of   1.273239545.

A Relationship between
Circles, Squares, Spheres and Cubes *

Let us go back, far back, to where we first saw the record of the circle's area being 28.2743388 and now let that value be the sphere's surface area. Note, that this number divided by nine, 9, **equals pi,** π, so the area of any cube's side will equal 1.5(π). Here, the radius of the sphere will be 1.5. The corresponding cube will have a side length of 2.170903764 and a volume of 10.2296744 because 28.2743388 is the circle's area being the sphere's surface area and that divided by six, 28.2743388/6, equals any area of a side of the cube, and $(28.2743388/6)^{0.5}$ = the length of any side of the cube which is 2.170803952. So, the value of the cub's volume must be 2.170803952^3 or 10.2296744. The volume of the sphere will be

$$4\pi r^2 = 4\pi 1.5^2 = 28.27433388.$$ Next, the volume of the sphere divided by the volume of a cube will be

$$\frac{28.2743338}{10.2296744} = 2.763952476$$

This number, 2.763952476, divided by two equals, don't act like you don't know what it is, 1.381976238, our "friend" from the relationship of a sphere and a cube, "The Second Very Important Number." Also, "this" number was divided by two because the original circle and sphere had a 3, but the last radius of the sphere had a 1.5 radius, which is three divided by two.

Some examples

1.) A circle has a radius of sixteen feet
    A.) The circumference is $2\pi r$ or   (100.5309649)[r/2] = 1.273239545
    B.) The area of a corresponding square is $[2\pi r/4]^2$ or   (631.6546817)

2.) A sphere has a radius of sixteen
    A.) the volume is $(4\pi r^3)/3$ = 17,157.28468
    B.) the circumference is $(4\pi r^2)$ =   3,216.990877
    C.) A corresponding cube will have a value for each of its sides of 804.2477193
    D.) The area of this cube will be 12,415.03272
    E.) $\dfrac{17,157.28468}{12,415.0732}$ =   1.381976598 our second important number.
        1.) $(0.004973592)(r^2)$ = 1.273239545, 4/π
        2.) Three times the radius times the area of the sphere divided by the area of the square is 1.273239545.

A Relationship between
Circles, Squares, Spheres and Cubes *

3.) A sphere has a radius of fifteen

        A.) the volume is $(4 \pi r^3)/3 =$ 14,137.16694

        B.) the circumference is $(4 \pi r^2) =$ 2,827.433388

        C.) A corresponding cube will have a value for each of its sides of
            471.238898.

        D.) The area of this cube will be 222,066.099.

        E.) $\dfrac{14{,}137.16694}{222{,}066.099}$ = 0.063661977

        F.) (0.028294212)(45 or three times the radius) = 1.273239545

        G.) $\dfrac{[2{,}827.433388]}{[222{,}066.099]}$ = $\dfrac{1.273239545}{100}$

Next! A circle has a circumference of nine feet, an area of 6.4455195 feet, and a radius of 1.432394488 feet. A corresponding square can be constructed from the materials of its circumference of nine feet. Here, each side of the square figure must include two and one quarter, 2.25, feet and the area of the square is $(2.25)^2$
Or 5.0625 square feet. The relationship of these two geometric figures is $4/\pi$ or 1.273239545.

$$1.273239545 \quad = \quad \frac{6.44577515}{5.0625}$$

The same relationship exists between spheres and cubes.

Next! Circle A is surrounded by Square A which is enclosed by Circle B. The relationship between the two circles is

$$\frac{\text{Area of Circle B}}{\text{Area of Circle A}} = 2.00$$

A Relationship between Circles, Squares, Spheres and Cubes *

$$\underline{\text{Area of Circle B}} \quad = \quad 2.0$$
$$\text{Area of Circle A}$$

If the smaller circle, Circle A, has a radius of seven, 7, the area of the circle is 153.93804 then the square's area is 14(14) or 196. Circle B's hypotenuse would be

$(14^2 + 14^2)^{0.5}$ = 19.79898987 and so the radius, R, would be 9.899494935 and the area would be 307.87608. The ratio of the two circles is
307.87609/ 153.93804 = TWO

The area of the square to the area of Circle A is 4/pi, 1.273239545,
Let's do another. The square has an X length of six, 6, and a Y length of six, 6. The area is thirty-six, 36. The hypotenuse, H, of the larger circle, Circle B, is $(6^2 + 6^2)^{0.5}$ which equals $(2)^{0.5}(6)$ which is 8.485281374 and the radius of this circle is 4.242640687. The area of Circle B is
$\pi\ r^2$ or 56.54866776.
Going further, circle A has a radius of three, 3, and since in a square the X dimension equals the Y dimension then $X^2 + Y^2$ is the same as $2X^2$ and we will set this to equal the radius squared or nine, 9.
$2X^2 = 9$ and then we see that X = $(4.5)^2$ or 2.121320344
This shows us that the area of Circle A is $(pi)(3)^2$ or 28.27433388 = TWO
This shows us that the area of Circle B is $2(pi)(3)^2$ or 56.54866776
We see that the ratio, again and like the previous situation, is two.

A Relationship between
Circles, Squares, Spheres and Cubes *

This process can be continued with adding more squares and circles to find more relationships. We can start by including another square around the bigger circle where the area of that square would be 9.899494935^2 or 97.99999997. The diagonal line through this square has a length of 14, and so on and so on and so on. Below there is an explanation a pattern of five square and circle combinations.

A Relationship between
 Circles, Squares, Spheres and Cubes *

In the below explanation there are five square/circle combinations. I will present them to you by giving you combinations, of increasing sizes, Combination A, Combination B, Combination C, and Combination D before I will present the and smallest of the combinations, which is Combination 1.

Combination A	Square A Area = 4 or 2^2 (Square 1)^2	Circle A Area $= \pi r^2$ $= \pi(1.414213562^2$ or $2^1)$ $= 2\pi$ Radius $= 1.414213562$ $= r$ Diagonal A $= 2r = 2.828427124 = $ (Circle A's Radius)^1 $\qquad\qquad\qquad\qquad\qquad\qquad$ = Circle 1's radius times 2 [(Diagonal A)/2](2.82842712) $=$ Diagonal B
Combination B	Square B Area = 8, or 2^3 Or (Square A)(2)^1 $= 2^3$	Circle B Area $= \pi r^2$ $= \pi(4$ or $2^2)$ $= 4\pi$ Radius $= 2$ $=$ Circle A's Radius)^2 Diagonal B$= 4$ $= d$ $= 2r$ (Diagonal B)/2(2.828427125) $=$ Diagonal C $\qquad\qquad\qquad\qquad\qquad\qquad\qquad$ (
Combination C	Square C Area = 16, or 2^4 (Square B)(2)^1 $= 2^4$	Circle C Area $= \pi r^2$ $= \pi(8$ or $2^3)$ $= 8\pi$ Radius $= 2.828427125$ $=$ Circle A's Radius)^3 Diagonal C $= 5.656954249$ $= 2r$ (Diagonal C)/2(2.828427125) $=$ Diagonal D
Combination D	Square D Area = 32 (Square C)(2)^1 $= 2^5$	Circle D Area $= \pi r^2$ $= \pi(16$ or $2^4)$ $= 16\pi$ Radius $= 4$ $=$ Circle A's Radius)^4 Diagonal D $= 8$ $= 2r$ (Diagonal D)/2(2.828427125) $=$ Diagonal E(Which is not shown)
Combination 1	Square 1 Area = 2 The length of each side of the square is 1.414213563	Circle 1 Area $= \pi r^2$ $= 1.50796326$ Radius $= 0.707106781$ Diagonal 1 = 1.414213562 $\quad$ The length of each side of the square is 1.414213563

Iyika

Olaniyan, in Philadelphia, PA, has discovered some fascinating relationships between circle lengths and areas. Here is his summary of a few of his findings. It is his express wish that these ideas be used to help many children find exciting math connections.

Iyika

This a Yoruba word that mean "circles" and it is the name of a method of measuring a circle's area, vertically or horizontally. The first term is of the Yoruba language, the second term is of the Swahili language, and the third term is of the English language. Here the X axis, or the independent variable, is along the circle's diameter line, which is labeled according to what percent of the diameter line's total length is every particular length on the diameter line from its far left point. The Y axis measures the percent of the circle that is below plus above the circle's diameter line and left of the vertical dividing line that separates that part of the circle that has been included from that part of the circle that has not been included. This vertical line can be moved to the left or to the right to decrease or to increase the amount of the circle's area that is included or accounted-for.

To find the circle's area percent of the circle's total area the circle is weighed. Then a new vertical strip of the circle is weighed and summed with other circle parts to show any usual or unusual patterns. Here we can find what percent of the total circle's weight is the new vertical strip of the circle. With that we can find the area, which must have a similar relationship to the total area of the circle as it has to the total weight of the circle.

As we move a vertical dividing line from the diameter's far left point the change in X is more than the change in Y until the fifty-fifth percent mark of X's length is reached where the Y percent is fifty-six and six thousand two hundred and sixty-five ten thousands. At fifty percent of line X's length we have the fifty percent of Y area but later the change in Y will be greater than the change in X and the percent of the Circle above plus below the diameter line and left of the vertical dividing line will be greater than fifty percent of the circle's area and the percent of X surpassed. This difference will increase until the eighty percent mark of the X axis is reached. From there the inequality will lesson until the one hundred degree mark of X and of Y is reached, where they will have equal values, one hundred percent, 100%.

There are variation of these facts. The facts depend on the size of the circle and the diameter of the circle.

EXAMPLE A

Letter	X	Y change	Y Cum	X %	Y Cum %	Y/X % or $(Y_{cum}\%/Y_{max})(100^2/X\%)$	
A	1	0.4	0.4	5	2.4096	48.0000	
B	2	0.5	0.9	10	5.4216	54.2000	
C	3	0.7	1.6	15	9.6385	64.2900	
D	4	0.8	2.4	20	14.4578	72.2891	
E	5	0.9	3.3	25	19.8795	79.5180	
F	6	0.9	4.2	30	25.3012	84.3373	
G	7	1	5.2	35	31.3253	89.5008	
H	8	1	6.2	40	37.3493	93.37325	
I	9	1	7.2	45	43.3734	96.3855	
J	10	1.1	8.3	50	50.0000	100.0000	
K	11	1.1	9.4	55	56.6265	102.9400	
L	12	1	10.4	60	62.6506	104.4100	
M	13	1	11.4	65	68.6746	105.6400	
N	14	1	12.4	70	74.6987	106.7000	
O	15	0.9	13.3	75	80.1204	106.8300	
P	16	0.9	14.2	80	85.5421	106.9270	
Q	17	0.8	15.0	85	90.3614	106.3400	
R	18	0.7	15.7	90	94.5783	105.0800	
S	19	0.5	16.2	95	97.5903	102.7200	
T	20	0.4	16.6	100	100.000	100.0000	

Note that $(y_{cum}/Y_{max})(100^2/X\%)$ equals any number of the far right column.

EXAMPLE B

Letter	X	Y change	Y Cum	X Cum %	Y Cum %	$(Y/X)\%$ or $(Y_{cum}\%/Y_{max})(100^2/X\%)$	
A	1	2.25	2.25	5	2.79		
B	2	3.06	5.31	10	6.5987		
C	3	3.44	8.31	15	10.8736		
D	4	3.89	12.64	20	15.7077		
E	5	4.51	17.15	25	21.3122		
F	6	4.31	26.26	30	26.6683		
G	7	4.83	26.29	35	32.6705		
H	8	4.93	31.22	40	38.7970		
I	9	4.99	36.21	45	44.9981		
J	10	5.23	41.44	50	51.4974		
K	11	5.06	46.50	55	57.7855		
L	12	4.81	51.31	60	63.7628		
M	13	4.64	55.95	65	69.5290		
N	14	4.75	60.70	70	75.4318	107.7597685	
O	15	4.38	65.08	75	80.8748	107.8331469	
P	16	4.23	69.31	80	86.1314	107.664347	
Q	17	4.07	73.38	85	91.1892	107.281486	
R	18	2.94	76.32	90	94.8427		
S	19	2.56	78.88	95	98.0241		
T	20	1.59	80.47	100	100.0000		

EXAMPLE C

Letter	X	Y change	Y cum	X Cum percent	Y Cum percent	Y/X percent or $(Y_{cum}\%/Y_{max})(100^2/X\%)$
A	1	10	10	5	2.1276	42.552
B	2	15	25	10	5.3191	53.191
C	3	20	45	15	9.5744	63.8293
D	4	25	70	20	14.8536	74.268
E	5	25	95	25	20.2127	80.8508
F	6	25	120	30	25.5319	85.10633333
G	7	30	150	35	31.9148	91.1851
H	8	30	175	40	37.2340	93.085
I	9	30	205	45	43.6170	96.9266
J	10	30	235	50	50.0000	100.0000
K	11	30	265	55	56.3829	102.5143636
L	12	30	295	60	62.7659	104.609833
M	13	30	325	65	69.1489	106.3829
N	14	25	350	70	74.4680	106.3828
O	15	25	375	75	79.7872	106.382933, $(375/470)(100^2/75)$
P	16	25	400	80	85.1063	106.382875
Q	17	25	425	85	90.4255	106.3829
R	18	20	445	90	95.7446	106.3828
S	19	15	460	95	97.8723	103.0234
T	20	10	470	100	100.0000	100.0000

EXAMPLE D

Letter	X	X Cum %	Change in Y	Y Cum	Y Cum %	$(Y/X)\%$ or $(Y_{cum}\%/Y_{max})(100^2/X\%)$	
A	1	5	0.051	0.051	1.7788	0.35576	
B	2	10	0.094	0.145	5.0575	0.3487	
C	3	15	0.114	0.259	9.0338	0.6022	
D	4	20	0.134	0.393	13.7077	0.6853	
E	5	25	0.152	0.545	19.0940	0.7637	
F	6	30	0.155	0.700	24.4192	0.8366	
G	7	35	0.174	0.874	30.4848	0.8709	
H	8	40	0.179	1.053	36.7282	0.9182	
I	9	45	0.191	1.244	43.3903	0.9642	
J	10	50	0.189	1.4335	50.0000	1.0000	
K	11	55	0.184	1.617	56.4352	1.02609	
L	12	60	0.178	1.795	62.6438	1.04406	
M	13	65	0.184	1.979	69.0617	1.0624	
N	14	70	0.168	2.147	74.9215	1.0703	
O	1	75	0.163	2.310	80.6069	1.0747	
P	16	80	0.157	2.467	86.0830	1.0752	
Q	17	85	0.140	2.607	90.9661	1.0701	
R	18	90	0.116	2.723	95.0122	1.0556	
S	19	95	0.093	2.816	98.2211	1.0339	
T	20	100	0.051	2.867	100.0000	1.0000	

Again, at eighty percent of X, Y is at it's greatest percent of X that is over one hundred percent. Here,

$$(Y_{cum}\%/Y_{max})(100^2/X\%) \text{ is the same as } (2.467/2.867)(100)(100)/(80) = 107.5601$$

Again, Duare or Ayika is a measurement of a circle's area, vertically or horizontally. Here, the X axis, or the independent variable, is along the circle's diameter line, which is labeled according to what percent of the diameter line's total length is a particular length on the diameter line from its far left point, and the Y axis measures the percent of the circle that is below plus above the circle's diameter line which also is on the left of the inclusion dividing line which is "a vertical dividing line." Here, inclusion represents that part of the circle that has been accounted for with a particular percent.

To find the area and the various circle area percent of the circle's total area the circle is weighed. Then a new vertical strip of the circle is weighed and summed with other circle parts to show any patterns.

As we move a vertical dividing line from the diameter's far left point towards the far right point the percent change in X is more than the percent change in Y, or the circle's area's size. This occurs until the fifty-fifth, 55, percent mark of X's length is reached where the Y percent is 56.6265. At fifty percent of line X's length we have fifty percent of the Y area but later the change in Y will be greater than the change in X and the percent of the total circle that is left of the vertical dividing line, while also being above plus below the diameter line, will be greater than fifty percent of the circle's area and it will be greater than the percent of "X" surpassed. This difference will increase until about the eighty or seventy-five percent mark of the X axis is reached. From there the inequality will lesson until the one hundred percent mark of X and Y are reached, where they will have equal values, 100%. Exceptions do occur that show when the growth in the area of the circle compared to the growth of X reaches its maximum difference.*

You should note some patterns of the Y Cum Percent column. In Example A, Row A divided by Row B is 2.4096/ 5.4216 or 1.00/2.225 and Row A divided by Row C is 2.4096/9.6385 = 0.249997406 or 2.50. There are other interesting results such as Row H divided by Row N of Example C being
37.234/74.468 – One-Half.
What I find most interesting is the fact that there is a lack of patterns one should expect to be there. Here, all circles are about the same whether the circles are small or large but the percent change in the amount of the circle included in the area above and below the diameter line and also left of the vertical dividing line, which can be moved from the left to the right or form the right to the left side of the circle, is not consistent. Why?

There are a few methods of constructing an equilateral triangle. One method is to use a compass and find three angles that all have the measurement of sixty degrees of an enclosed figure called a triangle.

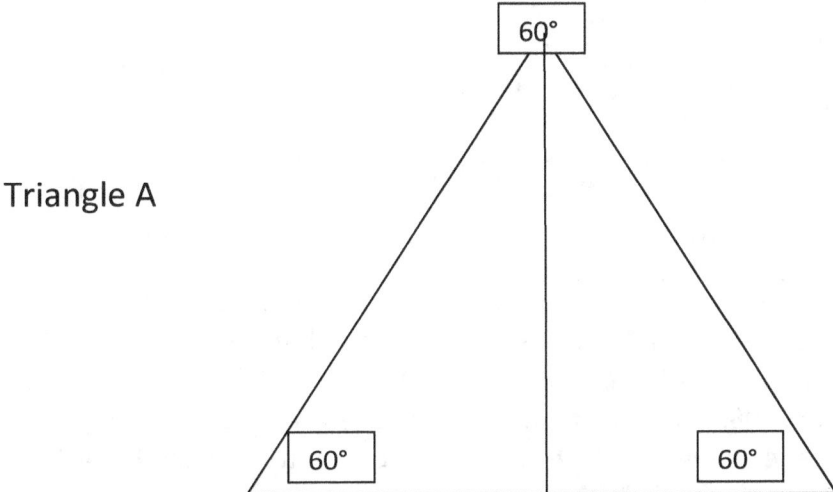

Triangle A

Another method of constructing an equilateral triangle is to allow one side of the triangle to be on the X axis and as you draw one of the other sides you give the slope of the other sides to have a three to the one-half power and a negative one times three to the one-half power slope, respectively. Also, the length of one of the sides divided by four-thirds to the one-half power or times 0.75 to the one-half power, where 0.75 equals $(1.333333)^{-1}$ will denote the height of the line that divides the triangle into two equal parts. The above facts of this paragraph is known as "The Law of Philly Signs." With this last fact comes the traditional sin of the angle times the hypotenuse equals the height of the triangle.

In the above triangle if each side has a length of ten inches then the height of the rise of the triangle will be ten divided by two and times a slope of three to the one-half power while the other side of the triangle will have a slope of negative one times three to the one-half power. The height of the triangle, the length of line "d," will thus be $(5)(3^{0.5})$ or 8.660254038 which is also the length of any of the triangle's sides divided by four-thirds which was raised to the one-half power, $10/(4/3)^{0.5}$, which is 8.660254038.

Again, this point can also be found by multiplying half of the length of a side of the triangle by the square root of the inverse of four-thirds, $10(3/4)^{0.5}$. Also, this place can be found by getting the sin of the 60° angle and multiplying it by the length of any of the sides of the equilateral triangle, (Sin 60°)(10). Note that the Cos of 60° times eight, 8, is 5, and this is where the axis of symmetry line meets the "x" axis.

The area of an equilateral triangle is $(S/2)^2(3)^{0.5}$.

What happens if the triangle is not an equilateral triangle? Can the above methods be used in places of this triangle?

Triangle B, like before, has a 60° angle on the right side and, here, has a twenty-inch hypotenuse for the triangle's right side and a sixteen-inch base that ranges below two hypotenuse lines. We can find the height of the triangle by getting sin 60° times twenty, 20. This will equal $10(3)^{0.5}$, or 17.32050808. This is like with the equilateral triangle WHERE the height of the triangle was found by dividing the sixty-degree angle hypotenuse line by $(4/3)^{0.5}$. This happened because, like before, the angle is 60° and this is a normal behavior of the sin function. The length of the horizontal line underneath the left hypotenuse line was found by solving for x in $x^2 + y^2 = h^2$.

$$\{6^2 + [(10)(3^{0.5})]^2\}^{0.5} = h \quad = \quad 336.^{0.5} = 18.33030278$$

Triangle C has a 40°, instead of a 60°, angle and has a twenty-inch hypotenuse line and a twenty-inch x line. So sin 40° (20) gives the triangle's height, which is 12.85575219, but twenty, 20, divided by $(4/3)^{0.5}$ does not equal this height. The x line portion below the twenty-inch hypotenuse line is 15.32088886 inches. This means that on the other side of this vertical line the x-axis has to be 4.679111138 inches long and the left hypotenuse line has to be 13.68080573 inches long while the inner longest line is 12.85575219 long. Again, in the last triangle the height is 12.85575219 inches, the right base side is 15.32088886, while the left base side is 4.679111138, and the left hypotenuse is 13.68080573. With the law of Cosines, the left angle is 70°.

307

THE LAW OF PHILLY SIGNS

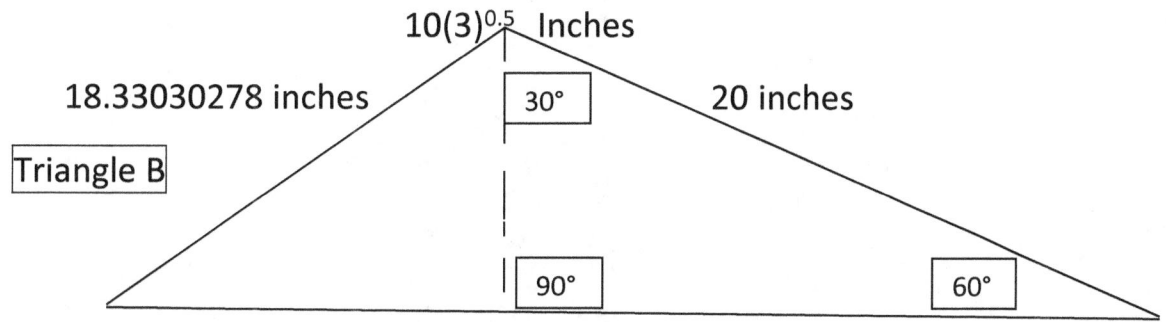

10(3)$^{0.5}$ Inches

18.33030278 inches

Triangle B

30°

20 inches

90°

60°

6 inches + 10 inches = 16 inches

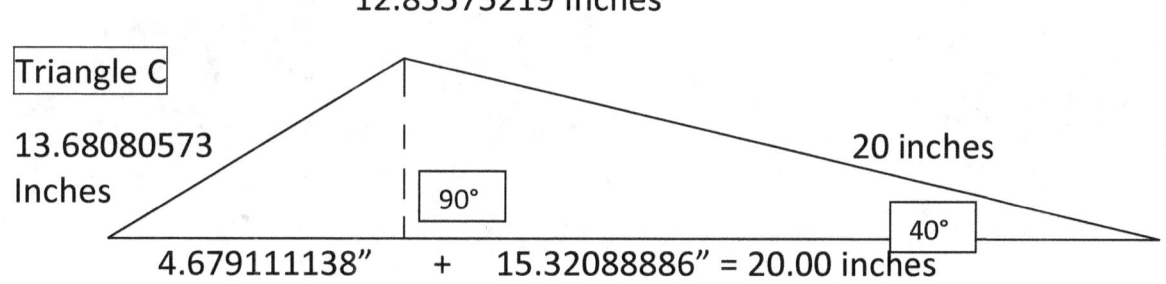

12.85575219 Inches

Triangle C

13.68080573
Inches

20 inches

90°

40°

4.679111138" + 15.32088886" = 20.00 inches

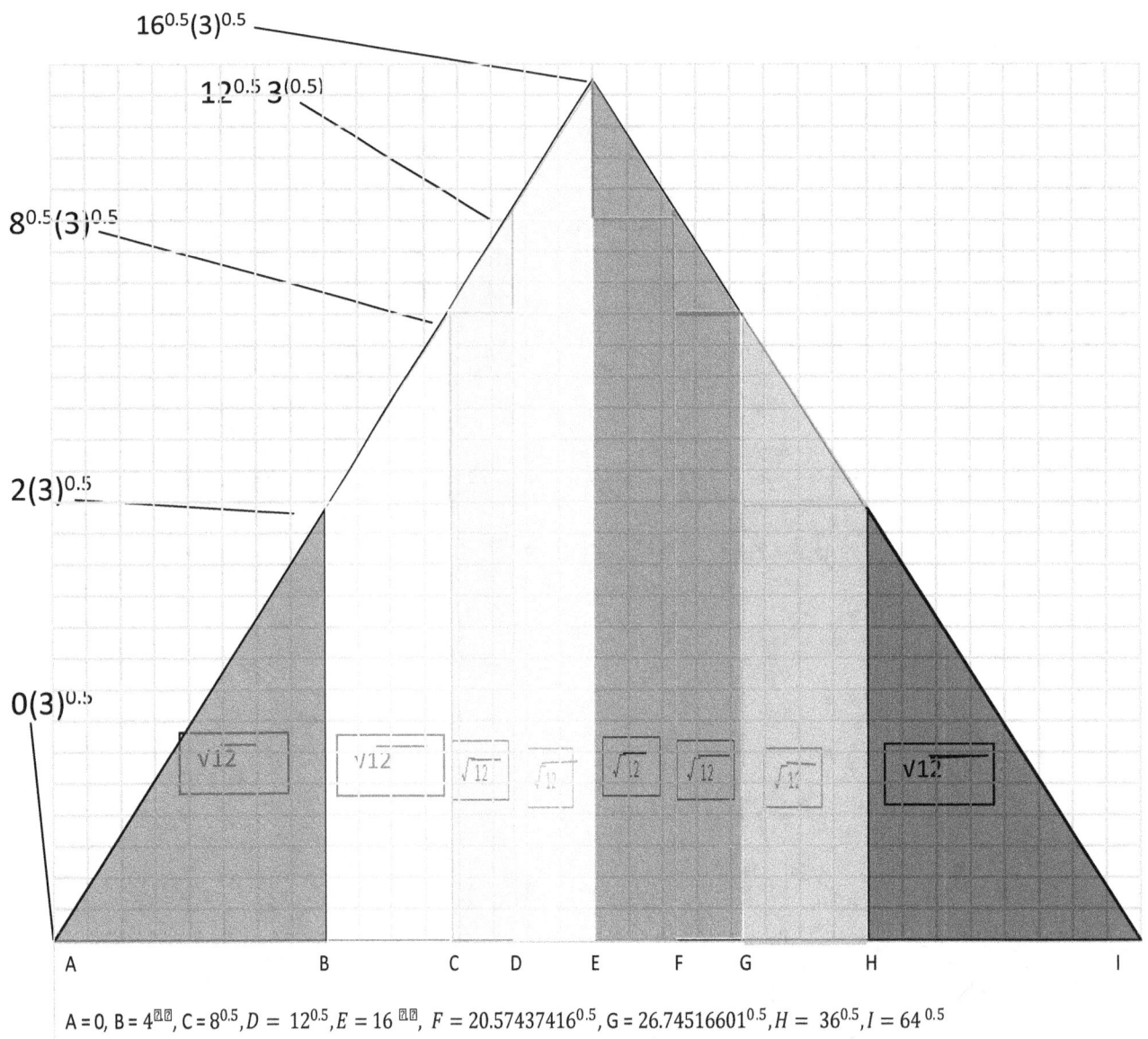

$16^{0.5}(3)^{0.5}$

$12^{0.5}\,3^{(0.5)}$

$8^{0.5}(3)^{0.5}$

$2(3)^{0.5}$

$0(3)^{0.5}$

A   B   C   D   E   F   G   H   I

A = 0, B = $4^{0.5}$, C = $8^{0.5}$, D = $12^{0.5}$, E = $16^{0.5}$, F = $20.57437416^{0.5}$, G = $26.745166601^{0.5}$, H = $36^{0.5}$, I = $64^{0.5}$

—The area of each of the seven colored sections is twelve to the one-half power square inches. The triangle's area is $16(3)^{0.5}$ or $(s/2)^2(3)^{0.5}$, where "s" is a triangle's side. The area of the above triangle equals $(3)^{0.5}(\text{width of } 8/2)^2$, where 8 is "s" and is the size of any of the triangle sides. This area, again, is 27.71281292 square inches.

309

About Our Brother's Place

This is a Shelter for males that is near Tenth and Hamilton Streets in Philadelphia, Pennsylvania. I stayed there before, sometimes during, and after I attending Cheyney University. I joined Back on My Feet's Jogging team when I lived there and after I stopped living there.

Notes

Notes

Index

Index

Index

Index

Index

Index

Index

Index

Index

Index

Index

A Country Home

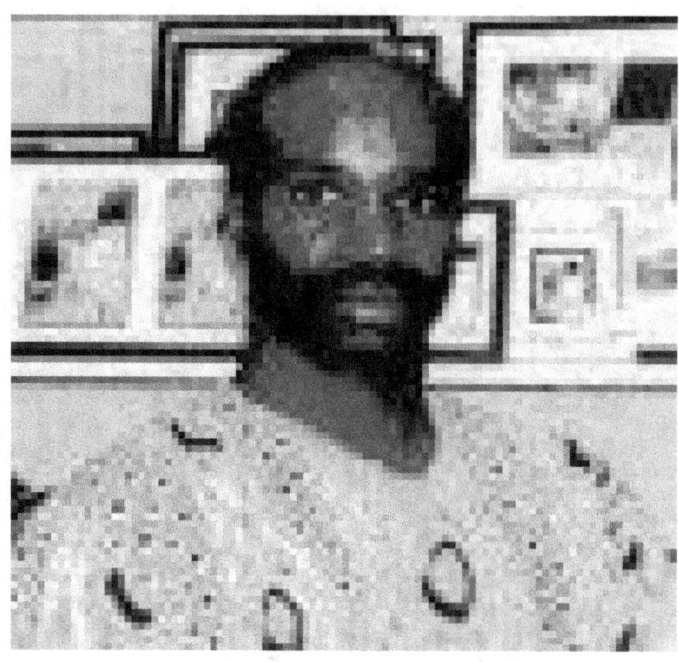

About the Author: Olaniyan Mtundu Adefumi was born in Philadelphia, Pennsylvania in 1958 in a neighborhood named "The Bottom." From there he attended Catholic schools before going to a Public school and before graduating from Cheyney University after majoring in biology.

In 1997 he was hit by an automobile as he attempted to bicycle from Canada to Mexico, within one month, to develop college scholarships for students. This coupled with issues of discrimination along the lines of race, religion, and perceived disability caused him to lose his work position at the Free Library of Philadelphia, Pennsylvania. Homeless as a result of being considered to be permanently disabled by a City of Philadelphia government doctor, he returned to school. There, in shelters such as Our Brother's Place, he noticed a lack of mathematical knowledge of many residents of shelters, so he wrote "Our Mathematics Book and Exponents for Our Brother's Place."

During his court case against the City, he attended college and graduated with honors. The Court found the City to be wrong and found him not to be disabled. From this position he is preparing to go to medical school so that he can soon operate as a medical doctor and develop college scholarships for students.

ISBN : 978-0-9778999-2-0